Y0-CQO-724

RESIDENTIAL MOBILITY AND PUBLIC POLICY

Volume 19, URBAN AFFAIRS ANNUAL REVIEWS

INTERNATIONAL EDITORIAL ADVISORY BOARD

ROBERT R. ALFORD
University of California, Santa Cruz

HOWARD S. BECKER
Northwestern University

BRIAN J. L. BERRY
Harvard University

ASA BRIGGS
Worcester College, Oxford University

JOHN W. DYCKMAN
University of Southern California

T. J. DENIS FAIR
University of Witwatersrand

SPERIDIAO FAISSOL
Brazilian Institute of Geography

JEAN GOTTMANN
Oxford University

SCOTT GREER
University of Wisconsin, Milwaukee

BERTRAM M. GROSS
Hunter College, City University of New York

PETER HALL
University of Reading, England

ROBERT J. HAVIGHURST
University of Chicago

EHCHI ISOMURA
Tokyo University

ELISABETH LICHTENBERGER
University of Vienna

M. I. LOGAN
Monash University

WILLIAM C. LORING
Center for Disease Control, Atlanta

AKIN L. MABOGUNJE
University of Ibadan

MARTIN MEYERSON
University of Pennsylvania

EDUARDO NEIRA-ALVA
CEPAL, Mexico City

ELINOR OSTROM
Indiana University

HARVEY S. PERLOFF
University of California, Los Angeles

P.J.O. SELF
London School of Economics and Political Science

WILBUR R. THOMPSON
Wayne State University and
Northwestern University

RESIDENTIAL MOBILITY AND PUBLIC POLICY

Edited by
W. A. V. CLARK
and
ERIC G. MOORE

Volume 19, URBAN AFFAIRS ANNUAL REVIEWS

SAGE PUBLICATIONS Beverly Hills London

Copyright © 1980 by Sage Publications, Inc.

All rights reserved. No part of this book may be reproduced or utilized in any form or by any means, electronic or mechanical, including photocopying, recording, or by any information storage and retrieval system, without permission in writing from the publisher.

For information address:

SAGE Publications, Inc.
275 South Beverly Drive
Beverly Hills, California 90212

SAGE Publications Ltd
28 Banner Street
London EC1Y 8QE, England

Printed in the United States of America

HT
108
.U7
vol. 19

Library of Congress Cataloging in Publication Data

Main entry under title:

Residential mobility and public policy.

(Urban affairs annual reviews ; v. 19)
Based on a conference held at the University of California, Los Angeles, November 12-14, 1979.
Bibliography: p.
1. Residential mobility–United States–Congresses. 2. Housing policy–United States–Congresses. 3. Policy sciences–Congresses.
I. Clark, William A. V. II. Moore, Eric G.
III. Series.
HT108.U7 vol. 19 [HD7293] 307.7'6s [307'.2'0973]
ISBN 0-8039-1447-4 80-12624
ISBN 0-8039-1448-2 (pbk.)

FIRST PRINTING

Contents

Preface 7

Part I: The Policy Context

1 □ The Policy Context for Mobility Research □
Eric G. Moore and W.A.V. Clark 10

2 □ Bringing Mobility Research to Bear on Public Policy □
Ira S. Lowry 29

3 □ Alternative Agenda in Academia and Public Policy □
Stephen Gale 34

4 □ Local Residential Mobility and Local Government Policy □
John M. Quigley 39

Part II: Research Contributions to Understanding Mobility 56

5 □ Demographic and Economic Change: Implications for Mobility □
James W. Hughes 59

6 □ Residential Mobility and Urban Policy: Some Sociological Considerations □
William Michelson 79

7 □ Housing Market Search: Information Constraints and Efficiency □
Terence R. Smith and W.A.V. Clark 100

8 □ Contemporary Housing Markets and Neighborhood Change □
James T. Little 126

Part III: Modeling the Impacts of Public Programs 150

9 □ Residential Mobility: Policy, Models, and Information □
Alan G. Wilson 153

10 □ Mobility and Housing Change: The Housing Allowance Demand Experiment □
Daniel H. Weinberg 168

11 □ Studying Residential Mobility: Administrative Records of the Housing Assistance Supply Experiment □
Mark David Menchik 194

Part IV: The Political Context 216

12 □ The Publiç City □
Michael Dear 219

13 □ Mobility, Community, and Participation: The American Way Out □
Norman I. Fainstein and Susan S. Fainstein 242

Part V: Informing the Local Decision Maker 263

14 □ Academic Research and Public Policy Formulation □
Martin E. Goldsmith and James S. Lemonides 266

15 □ The Public Policy Environment: Mobility Researchers' Responsibilities □
James E. Hartling 274

16 □ Data Resources for Monitoring Change □
Janet W. Byler and Randolf Gschwind 283

17 □ Continuing the Debate □
W.A.V. Clark and Eric G. Moore 308

The Contributors 317

Preface

□ THIS BOOK arises out of a conference entitled "Residential Mobility and Public Policy" held at the University of California, Los Angeles, on November 12-14, 1979. Experts from academic fields and practitioners involved in evaluation research and policy design at the federal level and individuals involved in program implementation at the local level attended the conference. Although all agreed that patterns of relocation of households in urban areas are of central importance in their policy work, they differed as to the nature of the problems of linking academic research and policy and, hence, of the intellectual paths to be pursued. The main purpose of this volume is to explore these differences and thereby provide a better understanding of the types of research on residential mobility which are likely to contribute to the policy process in the future.

This volume would not have been possible without the support of the Department of Housing and Urban Development's Office of Policy Development and Research, which provided generous financial support for the conference. In particular, we would like to thank Kathy Lyall (now of Johns Hopkins University) and Ken Weiand from HUD for their help in setting up the conference, and UCLA's Institute for Social Science Research for providing the support services necessary for the preparation of a manuscript.

–W.A.V.C.
–E.G.M.

Part I

The Policy Context

1

The Policy Context for Mobility Research

ERIC G. MOORE and W.A.V. CLARK

☐ MUCH CONTEMPORARY URBAN POLICY, particularly in the housing field, is concerned with providing incentives to or imposing constraints on household behavior in order to achieve policy goals. For example, attempts to improve the housing conditions of the poor contain complex sets of eligibility rules for obtaining loans or grants and efforts to stimulate neighborhood improvement range from increases in code enforcement activity to confining loan and grant eligibility to pecific areas. However, all such policies and programs face the same basic problem: The rules and procedures are laid down in an environment which is forever changing. Not only are the physical attributes of the cities in continual flux, but the composition, attitudes, and locations of the population are subject to constant adjustment.

Residential mobility, or the relocation of a household from one dwelling to another, is a fundamental element of the urban dynamic at both individual and aggregate levels. It is a consistent and pervasive behavior forming a major element of the policy context; it affects the conditions under which policies are developed and exerts a strong influence on their outcomes. However, although there is a voluminous literature on mobility, and much of the writing claims to have policy relevance, the contribution of this research to planning and public policy has been minimal.

The reasons for the poor links between academic research and policy formulation are complex. In part, the situation reflects the pressures to pose research questions within disciplinary frameworks which are motivated by issues quite different from those of making decisions and resolving problems within a policy arena. In part, academic research has been dependent upon the availability of data which are neither sufficiently up to date to address current concerns nor contain variables which can be related to the instruments of public policy such as tax incentives, housing allowances, building codes or zoning restrictions. More fundamentally, the limited contribution of research to policy reflects a failure to seriously examine the nature of the policy context and consider the demands on mobility research which arise from such a view. The primary function of this introductory chapter is to undertake such an examination and to suggest ways in which mobility research might contribute more readily to public policy.

THE POLICY CONTEXT

All too often, studies which claim to be aimed at policy begin with academic questions; having answered these questions within the framework of conventional social science, they then add the policy implications almost as an afterthought. However, it is important to recognize that the problems to which public policy is addressed are defined in the political arena, and unless research can be related to these political issues, then, indeed, it will have little to contribute.

Urban social programs and policies develop in highly complex circumstances. Only in part are they designed to ameliorate conditions perceived as problems by decision makers. The policies must also effect compromises between various jurisdictions and interest groups having different priorities regarding the allocation of public monies. Over time, programs and policies change in response to an evaluation of their relative successes and failures in achieving their goals, to changes in the underlying circumstances perceived as problems, to changes in the definition of what constitutes a problem, and to shifts in the distribution of power and priorities in allocating funds. Attempts to introduce analytic controls to cope with these changes abound in both theoretical and methodological work. The fundamental dynamism of the system is suppressed in models of residential structure which depend on assumptions of equilibrium, in surveys which examine conditions at a single point in time, and even in experimental designs which control initial conditions and strive heroically to keep them that way. The undeniable fact that the world will not stand still in an accommodating fashion poses continuing problems for the evaluation of public programs at all levels, but especially at the local level. This is particularly true for those programs concerned with housing and neighborhood improvement. As Quigley suggests in his chapter in this volume, equilibrium-oriented models are not very insightful when the main analytic questions concern changes in housing consumption in response to sociodemographic and economic events in a heterogeneous and uncertain market.

Those involved in designing and implementing public policy and associated programmatic decisions must face the reality of mobility and change directly. As the Fainsteins (Chapter 13) point out, mobility, whether social, occupational, or residential, is the accepted way of improving one's circumstances in the United States. Although there are constraints which apply differentially to social classes, particularly those relating to access to financial resources, society is organized to facilitate such patterns of adjustment. Even within western society there are major differences among nations. Mobility rates are significantly higher in the United States in comparison with European countries where state intervention in the housing market is far more direct.

The reconciliation of public programs with local dynamics raises a number of more specific issues. First, in designing policies and programs, the normal processes of mobility must be utilized if goals are to be attained. For example, in the majority of housing programs, consumption is not manipulated directly, but incentives are offered to induce participants to change their own housing consumption. Thus, it is important to understand the normal patterns of adjustment in order to ensure that programs do not merely replicate what would occur in the normal course of events. As a case in point, overcrowding has long been used as a measure of inadequate housing, yet a case study by Moore (1978) showed that, in one city (Wichita), over 30 percent of overcrowded households left that condition as a function of household or individual moves within 12 months while approximately the same number entered that condition. Any program to reduce overcrowding must take this dynamic property into account. A similar issue arises in considering current loan or grant programs for housing rehabilitation. Do they generate more rehabilitation, or do they merely substitute public monies for what might have been spent privately?

Second, policy design and implementation must recognize that continual adjustments in detail are required because exogenous events produce constant change in local conditions. Neighborhoods can exhibit substantial shifts in the amount of either abandonment or rehabilitation between one program year and the next, requiring adjustment in funding allocation. In part, the shift from the old categorical grant programs to community development block grants in the 1974 Housing Act was an attempt to give local authorities greater flexibility in designing and modifying programs to their own changing needs. However, if this flexibility is to be more than pure political reaction to local pressure groups, superior methods for monitoring and anticipating change are required. At present, our understanding of the dynamics of change, both in individual consumption and in the attitudes of neighborhood residents, is sufficiently limited that we must depend on direct observations, or monitoring, to tell us how things are changing. Such monitoring systems are required to perform the dual function of informing the local policy process and providing a basis for learning more about the effects of different local contexts (Gale, 1978).

Third, a corollary to the problem of program design and incremental change is the issue of evaluation. On the one hand, there are attempts to evaluate the impacts of systematic variation in program variables when many other factors which are not under program control are also varying; on the other hand, programs are introduced into situations which already have their own dynamic. In the latter case, questions of where that dynamic would have carried the system if the program had not been introduced are relevant. For certain types of programs, such as the demand experiment in the Experimental Housing Allowance Program (see Weinberg, Chapter 10), this problem can be addressed with adequate definition of test and control groups in the study design. However, for other programs, particularly those which are neighborhood-oriented, the difficulty of establishing either experimental or statistical controls is more serious.

At best, we might suggest that our analytical tools are not sufficiently refined to cope with dynamic systems, with the result that program effects usually turn out to be overwhelmed analytically by the "normal" patterns of mobility. At worst, we might suggest that the failure arises from asking the wrong questions, questions which stem from orthodox academic positions unsuited to the demands of informing decision-making and public action.

PEOPLE- VERSUS PLACE-ORIENTED PROGRAMS

In order to link research to public policy in an effective way, we must understand the nature of actions open to governments and then begin to trace the consequences of those actions. In western societies, although governments do have limited powers to directly induce mobility, their primary role is to influence the amounts of voluntary mobility by manipulating incentives, constraints, and contexts in which decisions are made.

(1) *The government can induce households to reappraise the relative benefits of staying or moving elsewhere by altering the attractiveness of their present residence.* Typical programs are those which provide loans or grants for rehabilitation of older dwellings, those providing housing allowances for designated households, and those promoting investment in specific neighborhoods leading to improvements in local amenities.

(2) *The government can directly influence the relative cost of alternate dwellings.* This can be achieved by legislative actions, such as the introduction of rent control, the use of various monetary or fiscal measures which alter effective mortgage interest rates, and the control of energy prices or the subsidization of public transit.

(3) *The government can change the cost of moving.* The most obvious action is the direct subsidy for moving costs, such as that given to those forced to relocate as a result of redevelopment. However, the cost of moving can also be affected by changing the cost of leaving the current residence. One example is the setting of capital gains taxes for owner-occupants. Another example, of a somewhat different type, occurs in the public sector in which criteria for allocation of public housing may include length of residence. Moving to another jurisdiction may then mean losing one's place in the queue. Both of these examples are transaction costs which may be sufficient to discourage mobility for a segment of the population.

A major problem for analysts is that these actions are highly interdependent. For example, rent control will often reduce costs for existing occupants, whose willingness to move is thereby lessened. However, fewer units then come onto the market, and the rents cannot be adjusted to accommodate the changing relations of demand and supply. Some alternative form of allocation must develop, such as the formation of queues administered by local authorities, payment of key money, or administrative allocation. Of course, whether it is more costly for households to wait in queues for rent-controlled housing or pay the market clearing rent is an open question, depending ultimately on where they are in the queue. Nevertheless, the control of rents will mean some households will be worse off by not being able to move when otherwise they could have done so.

These interdependent actions and their often counterintuitive outcomes are important considerations in evaluating current programs. For example, in current housing programs a distinction can be drawn between those programs which are primarily concerned with individual consumption and those whose main focus is on the characteristics of small areas or neighborhoods. (In some cases, the two perspectives are combined in complex eligibility rules which require that households receiving allowances, loans, or grants not only satisfy income and dwelling conditions but also live in designated neighborhoods; however, the thrust of the general argument is still valid). It has become clear that a concern of many city authorities has been to arrest the growth of blighted neighborhoods, to retain confidence in such areas and, hence, investment in the city. The end result (at least in theory) is to reduce the flow of more affluent households to suburban jurisdictions. This has led to changes in the framework of housing legislation, particularly in the Community Development Block Grant program, through which establishment of Neighborhood Strategy Areas as foci for investment of funds has introduced a stronger place orientation into programs whose major beneficiaries are intended to be low- and moderate-income households (Rosenthal, 1978).

An important issue arising out of this shift in emphasis is whether place-oriented programs can still fulfill their people-oriented goals of benefiting low- and moderate-income households. In part, of course, this is an empirical issue; but the question also has consequences for the way in which we think about mobility in a policy context. It is not just a question of developing better models of individual choice under new constraints, but of acquiring a deeper understanding of how society is structured to allocate goods, services, and valued resources. If we are just altering the location of desirable residential environments without changing any of the allocation rules, then the resulting transfer of value (desirable neighborhoods) to more powerful social groups in the form of gentrification is a predictable policy outcome. However, we must then also be concerned with the consequences of this redistribution; such consequences include a complex set of relocations and adjustments by former residents for whom assignment of benefits and costs in an overall evaluation equation is difficult. The point of this discussion is not that particular programs are right or wrong, but that we must understand both individual and social dynamics if we are to avoid specification of inconsistent goals and to appreciate what different programs are capable and not capable of achieving.

PUBLIC POLICY AND MOBILITY RESEARCH

Given this perspective on policy, how should we characterize existing research in mobility, and what are the promising directions in which research might proceed? Over the years, analysis of local residential mobility has been characterized by a strong split between studies of aggregate flows and studies of individual behavior. The former derive their primary impetus from studies of migration and of spatial interaction in general, while the latter have focused on economic and sociopsychological aspects of household decision-making (Clark and Avery, 1978). The intellectual origins of this work have led to emphasis on a somewhat narrow range of questions. At the macro level, the dominant concerns have been for identifying the structure of household differentials in the propensity to move and for representing flows between subareas in such a way that changes in the overall distribution of households can be derived. At the micro level, two basic themes have evolved. The first has attempted to build on the seminal contribution of Rossi (1955), in which the main goals have been to identify the nature of motivations and preferences underlying the decision to look for a new dwelling and to examine the structure of subsequent search, evaluation, and choice procedures. The second, reflected primarily in the recent work of Straszheim (1975), Hanushek and Quigley (1978), and Weinberg et al. (1979), has extended neoclassical location models of Alonzo, Muth, and Mills to the analysis of movers who enter the market in a given time period. Major distinctions exist between the behavior of the total population and the group recently in the market, perhaps best illustrated by analyses of the propensity to own (Struyk, 1976; Staszheim, 1975). It is only by making such distinctions that the recent problems of access to first-time ownership can be fully appreciated. The studies by Quigley and Straszheim stress the importance of adopting a dynamic perspective on the analysis of location decisions.

Most academic writers have treated the description and explanation of mobility behavior as the primary goals of analysis. However, in pursuing these goals, the emphasis has been on producing behavioral generalizations, and the importance of the local context has been downplayed. There is substantial empirical evidence on the contrasts in movement rates between sociodemographic groups and the general attitudinal correlates of the desire to move or stay (Speare et al., 1975). Yet, the studies concerned with processes such as urban renewal, office relocation, racial change, and the development of transportation systems must also include household mobility as an integral part of the definition of indirect and direct effects of public actions. The general literature has little to say on how this should be done.

The framework within which research questions are defined further limits their contribution to design, implementation, and evaluation of public policy. Two issues are of particular relevance. First, past research has tended to accept as legitimate the current socially determined frame of reference of planners and policy makers. Within this political context, there has generally been a poor understanding of how and why policy makers are concerned with residential mobility and the uses they are making (or not making) of the accepted research findings on residential mobility. The second limitation is the failure to recognize a significance to mobility beyond its relation to neighborhood change or individual housing consumption. In particular, there are broad-ranging distributional consequences of the continuing decentralization of the urban population, the most obvious example being the increasing division between black central cities and white suburbs. If we are to appreciate the potential of and the constraints on public programs, we must have a better understanding of the societal role of mobility (Harris and Moore, 1980). If we continue to place our faith in analytical studies of the individual without regard to the societal relations which define the context within which relocation adjustments are made, we are liable to design programs and policies whose outcomes fail to live up to our expectations.

Given these comments, it is possible to suggest a shift in emphasis in mobility research which might lead to a greater contribution to public policy. The subsequent discussion focuses on three main themes: the need to reevaluate our conceptualization of the nature of movement decisions, the need to provide a societal as well as an individual perspective on mobility, and the need to collect and organize data more appropriate to public policy analysis.

THE NATURE OF RELOCATION

Several changes in the way we view mobility could produce research results more readily translatable into policy statements. First, greater attention to outcomes rather than to the act of relocation itself would address a major criticism of existing work from a policy perspective. Second, a study of the linked nature of household decisions will provide a better understanding of the behavioral basis of mobility leading to more critical statements about responses to program incentives and constraints. Finally, the consideration of the joint behavior of housing and labor markets will lead to a sounder interpretation of the role of mobility in the broader urban system and of the potential impacts of changes in the local economy on residential redistribution.

Focus on Outcomes Rather than Actions

Much of the literature on mobility is concerned with the explanation of the act of moving, its frequency, timing, and sociodemographic context without regard to what is being achieved by the action, either individually or in the aggregate. The move itself is seldom, if ever, a concern of policy; however, it is important to know whether housing conditions are improved; whether accessibility to services or other centers of activity are enhanced; or whether greater flexibility in budgets is attained, neighborhoods integrated, or deterioration halted as a consequence of the aggregation of these individual acts. Thus, the documentation of mobility intention and outcome under different programs in varying market contexts is often of greater interest than attempting to explain why households moved in the past. This is especially true if the explanations are context-dependent, for they tend to provide only limited insights for current policy decisions. The shift in orientation means that the definition of outcome must derive from a consideration of what is meaningful in terms of program goals or, just as important, from fundamental notions of justice or equity. All too often, the approach to defining outcomes has reflected strong disciplinary perspectives, such as in the emphasis on distance and direction of moves by geographers, on changes in housing rents by economists, and on patterns of social relationships by sociologists. Unfortunately, these approaches seldom address the conditions which are of interest in formulating public programs; thus, it becomes difficult to relate such research to policy issues. For example, traditional efforts to evaluate the impact of actions, such as zoning, land-use control, tax policies, and even rehabilitation programs (Rothenberg, 1967), have focused specifically on the impacts on housing price, assuming that all shifts in household utility are adequately capitalized within the price mechanism. Clearly, this is a debatable issue both in regard to the efficiency of the market and to the appropriateness of measuring shifts in utility in a heterogeneous population through housing price. An evaluation of alternative definitions in terms of their contribution to policy decisions is critical in developing more incisive approaches to evaluation.

Links in Household Decisions

Once we start to define our questions in terms of the relations among process, outcome, and policy goals, we confront a second, more fundamental, issue. There are usually a variety of ways of achieving a given set of goals both from an individual and from a societal perspective. The decision to move, as Hanushek and Quigley (1978) point out, is an important way to adjust housing consumption, but it clearly is not the only way. Hirschman's (1974) application of the exit, voice, and loyalty concepts to residential relocation laid out a more general structure, which, in detail, turns out to be crucial to policy (Moore and Harris, 1979). For example, the choice between residential improvement and moving is a critical investment decision for a household. The decision of a spouse to enter the labor force may facilitate the move to suburban home ownership, while the decision to join a group lobbying for better local schools may stop a move to another school district. Public programs are not only influenced by these trade-offs, but sometimes, as in the case of rehabilitation programs, depend on such trade-offs for the goals to be achieved.

The Joint Behavior of Housing and Labor Markets

The importance of the linkage between decisions can be extended to an analysis of more general ideas about residential location and urban structure. For a long time, the journey to work has figured prominently in analyses of residential location. Although considerable skepticism has been expressed as to the utility of the equilibrium-oriented neoclassical models of urban structure, work-home relations are still of major concern from a number of different perspectives. First, continuing shifts of employment opportunities in urban areas have important consequences for relocation behavior of different groups. Perhaps one of the most important issues is to evaluate the effect of changing patterns of family structure and labor force participation on traditional analyses of housing consumption and locational choice. The growth of single-parent families, nonfamily households, and two-wage-earner families have major implications for locational decisions and consumption choices which have received scant attention to date. Classical analyses implicitly assume a stereotypical household for which the head's journey to work is of major importance in the location decision. Yet, we now find, for example, that the majority of households entering the owner market for the first time have two wage earners; the impact of this situation both on initial location and subsequent housing experience is poorly understood, as is the experience of those who do not have such multiple resources.

Like access to jobs, access to services largely involves journeys from the home. However, the policy implications of these journeys to services are somewhat different and may be more susceptible to political decision-making. Many of the questions relating to the delivery of services have been modeled in a location-allocation framework, without a detailed consideration of changes in the population composition and distribution within the city. In other words, the consequences of residential mobility have been largely ignored. Residential mobility has often invalidated the locational decisions of health, educational, and other community agencies, yet little attention has been given to the implications of these effects for future decisions.

THE SOCIAL AND POLITICAL ROLE OF MOBILITY[1]

Since Tiebout's (1956) argument that movers vote with their feet, in the sense of choosing between locationality-dependent bundles of public goods, there has been a real awareness of the political consequences of movement behavior. These political consequences are not easily separable from issues of redistribution, access to services, and the links between public programs and mobility. Recent discussions of the role of institutions have further emphasized the need for a better understanding of the way in which the structure of society, particularly the political structure, influences differential access to segments of the urban system (Masotti and Lineberry, 1976). Not only are direct links important, such as those between mobility and the changing tax base in cities and between white flight from the central cities and quality of urban education (for which the political implications are clear), but we must also consider more subtle relocation effects, such as those arising from eligibility rules for loans and grants and possible discriminatory effects of neighborhood-oriented programs.

Within the larger social and political context we are also led to ask a different set of questions. Not only are we concerned with who moves and why they move (Simmons, 1968) but also the following: "How does the larger social system mediate the adjustments of different groups to needs and opportunities?" "What effects does mobility then have upon that social system?" "To what extent does the societal structure itself create differential needs for housing adjustment?" In addressing these broader questions we are faced with the problem of developing an understanding of individual behavior which is compatible with theories of social change.

Some important advances have been made recently in this regard. A considerable body of research has grown up which discusses the way in which household decisions are constrained, mediated, and satisfied by a variety of groups and institutions operating in the housing market (Murie et al., 1976; Duncan, 1976; Harvey, 1974). There has been a tendency for this work to be one-sided. It has addressed the questions of how housing needs and moves are created and mediated, but has remained almost silent on the question of subsequent effects of household movement. These effects are felt not only by the household but also by the larger social system. Neither effect has received adequate treatment; as a result, these relations between individual and societal effects should constitute a joint focus for future research.

The problem here is that of conceptualizing the reciprocal relationship between the individual household and society. We must recognize that the way needs arise for different groups and the way adjustments are made reflect quite specific social conditions and attitudes. Unfortunately, a cross-sectional analysis which focuses on consumption before and after a move is insufficient, although clearly a popular strategy in attempts to define the costs and benefits of moving (Pack, 1973). Although there is a great difference, for example, in the interpretation of public programs which facilitate entry into new housing markets as opposed to programs which merely advance the time at which entry is made, cross-sectional analysis cannot distinguish between the two types of effects. Similarly, a move is merely an initiation of a complex set of adjustments of activities, resources, and budget allocations in a new environment, and initial consumption is not necessarily a good indicator of these impacts. Recent work by Michelson (1977) on the relationship between life chances and residential choice in general and by Golant (1977) on the elderly represent the types of specific studies which follow logically from our earlier discussion.

At scales greater than that of the household, cross-sectional analyses are also of very limited value. This has been recognized in recent arbitrage models of neighborhood change (Leven et al., 1976). Such models are among the first to treat explicitly the dynamic and reciprocal relationship between individual choice and institutional behavior. Beyond the neighborhood (apart from some broad generalizations regarding the recent effects of the suburbanization of white, middle-class families), the specific social and political impacts of intraurban mobility have received scant attention. Where social and political impacts are discussed—and this is the critical point—they are not explicitly interpreted in relation to the specific needs of the household or to the particularity of the locations in question.

The need, then, is for longer-term studies covering a number of years. Changes in patterns and types of movement may then be interpreted within the context of broader shifts in the creation, accumulation, and distribution of wealth and of the development of housing markets and housing policy. Such studies must take account of the fact that, while mobility is in one sense a response to social change, it also has effects on that wider process.

INFORMATIONAL NEEDS

One of the more important implications of the earlier discussion of the policy context is that there is often a dearth of knowledge about local circumstances and the way in which situations change as programs are implemented. Decisions are still made even when theory is poorly articulated and when understanding of the consequences of actions is weak. A need exists to build up a system of data collection which serves both public policy and the development of knowledge.

There is strong evidence that the patterns of relocation and housing consumption vary substantially among cities (eg. Struyk, 1979). This variability leads us to suggest that transferability of existing decision-making models of mobility and housing adjustment from one local context to another may be very limited. In part, the reason lies in the fact that current models are almost entirely dependent on the theory of demand; in a market in equilibrium, characteristics of supply are of little consequence. However, if we consider consumption in a single city, in the short run, to be a function of relatively inelastic local supply, then models which only include demand-side variables must be seriously misspecified. Model parameters will not only vary across metropolitan contexts but will also be difficult to interpret. Clearly, we can seek more complex models which include additional variables to represent differences in local context, but whether they will prove to be estimable given available resources is debatable.

Whether or not we are able to make use of models, local data on housing market attributes, population composition, and mobility are necessary to support local decision-making, to provide a context for program design, and to chart the path taken by the local system under the existing set of institutional rules. What emerges as a critical need are strategies for local data collection and organization capable of performing these decision-support functions. This is a complex issue, since local communities have only limited resources for collection of new data for policy analysis; one answer seems to lie in more effective use of existing data files constructed for administrative and management purposes, although this usage poses additional problems of confidentiality.

There is substantial interaction among data, theory, and practice. Data are not neutral. For example, the strong emphasis on equilibrium models of urban residential structure has certainly fed on the availability of census data as decennial cross-sectional samples. Once detailed micro-level data became available, issues of market heterogeneity surfaced on a more persistent basis (Straszheim, 1975). The recent construction of longitudinal files poses new pressures to recognize disequilibrium states as the norm. The implication is that we cannot afford to reject local administratively generated data as inadequate. The policy process demands detailed, up-to-date monitoring of local conditions, and administrative data are often the only source which comes close to satisfying the need.

THE STRUCTURE OF THIS VOLUME

The chapters in this volume are organized in such a way as to illustrate the main ideas contained in this introductory essay. Clearly, there are differences in opinion among the authors—substantively, methodologically, and ideologically. That is the essence of a research agenda on such a broad-ranging topic. However, each of the essays is concerned with advancing our ability to relate academic research to public policy.

The book is divided into five sections representing different ways of linking mobility research and public policy. In the introductory section, the stage is set in terms of the nature of the public policy process and the general requirements imposed by this process on academic research. In Section II, four chapters extend existing social science perspectives on mobility in directions which have important implications for policy. The nature of changes in family structure, of conditions which either force households to move or prevent them from moving, and of the role of information in housing search are explored and their impact on policy discussed. Apart from their innovative contribution to the understanding of mobility, they are linked by a general style which emphasizes relocation as a social phenomenon that is relevant to policy but that does not include instrumental variables within the analysis. This position is taken a step further in Section III, in which models of mobility behavior with intellectual origins clearly in academic research are applied to policy evaluation. These models are constructed within the specific context of the U.S. Department of Housing and Urban Development's Experimental Housing Allowance Program. Both empirical pieces take the logical and necessary step of including instrumental variables, such as the structure of allowance incentives or eligibility rules, directly within the framework of analysis.

Sections II and III provide a largely orthodox, neoclassical perspective on mobility analysis in which theories of individual behavior dominate. However, if we are to understand the more general role played by mobility in society and hence the structural context within which policies can be pursued, we must broaden our discussion. In Section IV two chapters are presented which consider, first, the general locational response of a broad range of governmental programs aimed at providing services to disadvantaged groups and, second, the potential strengths and limitations of community action in promoting neighborhood stability. The common ground is that both present their arguments within the broad framework stimulated by the Marxist theory of urban structure.

Section V presents a quite different viewpoint on the link between mobility and policy. From the standpoint of the practitioner, the urban planner, or the decision maker, the primary need is to respond to current problems as defined within the local or even national political context. However, local conditions change rapidly, a situation clearly relevant to our discussion of mobility. We must recognize that our understanding of this phenomenon is not sufficient to be able to predict either the path of change or the adjustments induced by particular programs at the level of detail demanded by decision makers. A premium is therefore placed on the ability to monitor both the process of adjustment and the resulting distributional changes. The emphasis is on the design of data collection systems which are within the capabilities of local government but which must be structured in light of existing knowledge of the variables important to mobility. This need to inform local decision-making thus forms the focus of Section V. The final essay reviews the contributions contained in the volume in light of this introductory statement and of the discussions generated in the November conference.

NOTE

1. This section draws on a similar argument in Harris and Moore (1980).

REFERENCES

CLARK, W.A.V. and K. AVERY (1978) "Patterns of migration: a macroanalytic case study," pp. 135-196 in D. Herbert and R. J. Johnston (eds.) Geography and the Urban Environment. London: John Wiley.

CLARK, W.A.V. and E. G. MOORE [eds.] (1978) Population Mobility and Residential Change. Studies in Geography 25. Evanston, IL: Northwestern University.

DUNCAN, S. S. (1976) "Research directions in social geography: housing opportunities and constraints." Transactions of the Institute of British Geography: housing opportunities and constraints." Transactions of the Institute of British Geographers. New Series 1: 10-19.

GALE, S. (1978) "Remarks on information needs for the study of geographic mobility," pp. 13-48 in W.A.V. Clark and E. G. Moore (eds.) Population Mobility and Residential Change. Studies in Geography 25. Evanston, IL: Northwestern University.

GOLANT, S. M. (1977) "Spatial context of residential moves by elderly persons." International Journal of Aging and Human Development 8: 279-289.

HANUSHEK, E. A. and J. QUIGLEY (1978) "Housing market disequilibrium and residential mobility," pp. 51-98 in W.A.V. Clark and E. G. Moore (eds.) Population Mobility and Residential Change. Studies in Geography 25. Evanston, IL: Northwestern University.

HARRIS, R. and E. G. MOORE (1980) "An historical approach to residential mobility." Professional Geographer 32: 22-29.

HARVEY, D. (1974) "Class-monopoly rent, finance capital and the urban revolution." Regional Studies 8: 239-255.

HIRSCHMAN, A. O. (1974) "Exit, voice and loyalty: further reflections and a survey of recent contributions." Social Science Information 13: 7-26.

LEVEN, C. L. et al. (1976) Neighborhood Change: Lessons in the Dynamics of Urban Decay. New York: Praeger.

MASOTTI, L. H. and R. L. LINEBERRY [eds.] (1976) The New Urban Politics. Cambridge, MA: Ballinger.

MICHELSON, W. (1977) Environmental Choice, Human Behavior and Residential Satisfaction. New York: Oxford University Press.

MOORE, E. G. (1978) "The impact of residential mobility on population characteristics at the neighborhood level," pp. 151-181 in W.A.V. Clark and E. G. Moore (eds.) Population Mobility and Residential Change. Studies in Geography 25. Evanston, Illinois, Northwestern University.

--- and R. HARRIS (1979) "Residential mobility and public policy." Geographical Analysis 11: 175-183.

MURIE, A., P. NINER, and C. WATSON (1976) Housing Policy and the Housing System. Urban Regional Studies No. 7. London: George All en & Unwin.

PACK, J. R. (1973) Household Relocation: The New Haven Experience. Philadelphia: University of Pennsylvania Institute of Public Policy Analysis.

ROSENTHAL, D. B. (1978) "Neighborhood strategy areas: HUD's new initiative in neighborhood revitalization." Journal of Housing 35: 120-121.

ROSSI, P. H. (1955) Why Families Move. New York: Free Press.

ROTHENBERG, J. (1967) Economic Evaluation of Urban Renewal. Washington, DC: Brookings Institute.

SIMMONS, J. W. (1968) "Changing residence in the city: a review of intra-urban mobility." Geographical Review 58: 622-651.

SPEARE, A., S. GOLDSTEIN, and W. FREY (1975) Residential Mobility, Migration and Metropolitan Change. Cambridge, MA: Ballinger.

STRASZHEIM, M. H. (1975) An Economic Analysis of the Urban Housing Market. New York: National Bureau of Economic Research.

STRUYK, R. J. (1979) "The need for local flexibility in U.S. housing policy." Policy Analysis 3: 471-483.

--- (1976) Urban Homeownership. Lexington, MA: D. C. Heath.

TIEBOUT, C. M. (1956) "A pure theory of local expenditures." Journal of Political Economy 64: 416-424.

WEINBERG, D. H., J. FRIEDMAN, and S. K. MAYO (1979) "A disequilibrium model of housing search and residential mobility." Presented at a conference on Housing Choices of Low Income Families, Washington, D.C.

2

Bringing Mobility Research to Bear on Public Policy

IRA S. LOWRY

☐ IN THESE NOTES, I propose first to characterize the policy issues to which research on mobility is plausibly relevant. Then I will try to distinguish the ways in which mobility research may be brought to bear on them. I conclude that, whatever the general scientific virtue of mobility research, policy today is served mostly by supplying organized information to inform public discussion.

THE POLICY ISSUES

Three mobility-related issues have engaged the national attention for the past quarter-century.

INTERREGIONAL REDISTRIBUTION OF POPULATION

After World War II there was a dramatic movement of southern blacks to northern and western cities, a movement of Puerto Ricans to the northeast, and a steady flow of immigrant Latins to the southwest. More recently, the pattern of flows has significantly altered, the important point being that the industrial "snowbelt" is emptying. These movements have been issues of national concern, primarily because they create fiscal problems for local jurisdictions. Inmigration creates demands for infrastructure, and inmigration by poor people creates demands for transfer payments and social services. Outmigration reduces local tax revenues and characteristically leaves behind the elderly, the less skilled, and the less able.

URBAN FORM

At the local level, the past quarter-century has seen a powerful dispersive movement from the peaks of urban density, leaving, in many cases, a half-empty core of abandoned factories, homes, and retail stores. This bleak central-city landscape offends our sense of fitness—our intuitive feeling that in a better world cities would grow old gracefully. It has also entailed a redistribution of wealth between owners of central-city property and those who stockpiled peripheral land. Finally, the erosion of central-city tax bases has made their governments dependent on federal subventions for their survival.

RESIDENTIAL SEGREGATION AND INTEGRATION

A quarter-century ago, the federal government, led by the Supreme Court, launched a policy of racial integration of a highly segregated social system. A general premise of this integration policy is that the opportunity set of the low-status groups can be altered by federal fiat, and that as the aided groups take advantage of these opportunities, their social status will improve until, eventually, the issue of segregation disappears. There have been dramatic strides in employment integration, substantial ones in school integration, and hardly any at all in residential integration. It seems clear now that if the integrative policy is to be pursued to fulfillment, it will require either much more draconian measures than have so far been applied or a much longer time than its enthusiasts care to acknowledge.

To be policy-relevant in an important way, residential mobility research must bear on at least one of these three issues. I hasten to add that research that does not meet that test may still serve useful purposes at a programmatic level. For example, an urban renewal administrator may need advice on relocation problems, or a welfare administrator may need to know the probable growth in his caseload next year. Research need not always be justified by its policy-relevance, except in an application for federal funding.

HOW RESEARCH CAN INFLUENCE POLICY

The first thing to note about the issues I have described is that they concern the results of mobility, not the process itself. Changing one's residence is considered in our society to be a natural right, not subject to question, and for each individual mover the move is an episode. Collectively, however, the moves affect the interregional distribution of population, urban form, and the racial mixture of neighborhoods. Manifestly, understanding mobility will teach us something about its static consequences.

There are several ways in which mobility research can be brought to bear on the big policy issues.

It can tell us what is happening to those issues. For instance, we first learned of the reversal of interregional population flows from the Census Bureau—not by static enumeration but by their annual estimates of the interregional migratory flows. In a sense, this descriptive function is served mostly by systematic data-gathering rather than by analysis. But every data-collection program has an implicit theoretical structure that defines the unit of observation and the characteristics of that unit that are worth recording. Research into the causes of residential mobility can call attention to failures in existing statistical systems to attend to policy-relevant aspects of mobility, suggest appropriate levels of spatial aggregation, or specify the useful dimensions of cross-tabulation.

Research can inform our choices of outcomes to be sought. Is it a good thing that the snowbelt is now emptying into the sunbelt? Is "spread city," on the whole, better or worse than "beehive city"? What type of residential integration best serves the elimination of racially based status distinctions?

Social scientists can offer a few principles as guides to better or worse. For example, the economist's efficiency principle tells us that we are collectively better off if capital and labor both move about until their marginal products are everywhere equal.

But such principles are scanty, and, in any case, balancing incommensurable goals remains a political process. What research can do is to illuminate the consequences of past choices and, sometimes, predict the consequences of choices still to be made, with special attention to side effects likely to be overlooked by those of narrower vision.

Finally, given a policy, mobility research can evaluate alternative instruments. If we can agree that revitalizing central cities is a national priority, how do we persuade people to reconcentrate residentially? Will housing subsidies be effective? Improvements in public services? Job creation? Designing specialized neighborhoods to recruit residents selectively? Raising the cost of travel?

Evaluating alternative policy instruments seems to me an especially important function. Consultation work with public agencies has taught me that an agency's chosen policy instruments often have only tenuous connection with that agency's expressed policy goals. In particular, policy instruments in the fields under discussion here are nearly always designed to produce a desired local effect; the assessment of outcomes rarely uses globally closed accounts. In consequence, robbing Peter to pay Paul is the norm, not the exception.

WHAT MODES OF RESEARCH ARE HELPFUL?

I have observed useful research results emerging from a wide variety of research styles. Serendipity is at least as important as good research design or rigorous methodology. However, I would like to close by commenting on two issues. First, there is a dangerous tendency among researchers to reify complex models, especially so-called simulation models. These are built on the premise that the mysteries of system behavior are mysteries because the interactions are too numerous to summarize in a closed-form analysis. A computer is used as an indefatigable bookkeeper. However, the existence of all this bookkeeping machinery tends to lull the investigator into forgetfulness. The description of initial conditions and the estimates of flow parameters are typically crude in a large interacting system; errors are propagated and, under many circumstances, magnified. In my judgment, the proper use of the large simulation model is heuristic, not for predictions.

Second, to influence policy, research results must be translated into the vocabulary of the intended audience. There are various audiences that have a role in policy formulation–legislators, executive agencies, advisory commissions, interest groups of influentials outside government, the academic community, and the voting public. Each has a somewhat different method of formulating research results and a different capacity for accepting abstractions. With the exception of academics, subtle analytical distinctions and marginal empirical differences rarely change an audience's prior views or give useful support to an intended policy initiative.

3

Alternative Agenda in Academia and Public Policy

STEPHEN GALE

□ RESIDENTIAL MOBILITY poses a problem for the analyst not so much because it is complex conceptually, but rather because it is a pervasive phenomenon, impinging on many issues but central to few. Although the intellectual lore of the mobility research of the past half-century has emphasized the singular efficacy of explanatory questions such as "Why do families move?", it is now becoming increasingly clear that such inquiries are, if not misplaced, certainly of limited value. Of course, it is true that the explanation of mobility is complicated, but so is the explanation of any other social phenomenon which has economic, demographic, psychological, geographic, and other influences. And, of course, it is true that a general explanation of the mobility process would help set the guidelines for effective public action. However, at least from a pragmatic perspective, what is interesting and important about the contemporary study of mobility is that it is at once a cause, an effect, and an intermediary; localized and spatially extended; a consequence of public and private policies and a policy instrument; endogenous and exogenous. The focus and contexts of mobility research are clearly heterogeneous, and it is this heterogeneity, I claim, which is characteristic of the academic and policy analysis agenda.

As I am sure we all recall, the two decades immediately following World War II were a time of visionary science; a vision of complete, consistent social theories which would embody both positive and normative perspectives. Social psychology, in particular, was pictured as playing a central role in the development of such theories. Questions such as "Why do families move?" were effectively transformed into requests for information and models of individual and group behavior in complex settings. Families, as groups, were characterized in terms of a variety of economic, social, demographic, and related attributes; within particular environmental contexts, families were then viewed as assessing their utility for place in light of jobs, familial needs, housing, and so on. The anticipated theory was to have had at least two parts: one which assessed the degrees to which various factors influenced the choice of place, and a second which prescribed locational changes in light of comparisons with some normative base. At the least, the vision held out an expectation of knowledge; in principle, this knowledge was also to provide for an immediate translation into the kinds of prescriptions social scientists like to call "policies."

It takes little post hoc vision—or, indeed, understanding of the development of the social sciences in the past twenty years—to see that much of the hope of these early years was rather ill-founded. The vision of a "grand" theory was clearly premature; perhaps most importantly, the expected translation of the positive into the normative was beyond even conceptual reach. Understanding and the grounds for understanding is one thing; but the grounds for translating understanding into prescription and action were obviously qualitatively different.

To a certain extent, my arguments on these issues have been effectively made in an earlier paper in a similar forum (Gale, 1978); in the interests of clarity, however, it is worth reiterating the central points here. My claim was very simple: Since the policy process consisted of a variety of types of questions (that is, requests for information and decisions), the easiest way to conceive of the heterogeneity of mobility research is by indicating the multiple types of information requirements for answering particular questions. A typology was developed to integrate questions, activities, and strategies (Gale, 1978: 35). Types of questions are, in effect, matched with particular types of information, and, barring the development of a comprehensive theory and measurement system, the typology implied a considerable heterogeneity in the questions and analytic strategies associated with the study of residential mobility.

What was perhaps not apparent from my earlier argument is the rather oblique way in which research on mobility enters into most academic and policy studies. Consider, for example, three general types of studies which, in one way or another, involve the analysis of patterns of residential mobility: disciplinary studies, ideological inquiries, and pragmatic problem-solving. Consider the typical sets of questions arising in each.

DISCIPLINARY STUDIES

Sociology/demography. Why do families move? How does mobility affect natality, fertility, mortality? How does residential mobility affect the stability of communities?

Economics. How does residential mobility affect economic equilibrating processes? How does the locational distribution of employment and infrastructure affect the mobility process? What are the effects of economic incentives on relocation decisions?

Geography. What is the spatial distribution of movers and nonmovers? Is there a distance and/or directional bias in locational decision-making? How does the perception of distance affect the decision to relocate? How sensitive are decision makers to the characteristics of local areas rather than dwellings?

Political science. How does mobility affect political decision-making? Does a highly mobile society require new forms of political organization?

IDEOLOGICAL INQUIRIES

Neoclassical economics. Do observed mobility decisions reflect the operation of market-based decision procedures? Can mobility patterns be affected by adjustments in the regulation of housing markets?

Marxism. How does mobility affect social equality? How does mobility affect the processes of labor market and residential segmentation? To what extent does the pattern of relocation reflect the constraints of the larger class-dominated social structure as opposed to the preferences of individuals?

4

Local Residential Mobility and Local Government Policy

JOHN M. QUIGLEY

☐ AN UNDERSTANDING OF THE FACTORS influencing household mobility over space is of considerable importance in the design and evaluation of a broad range of economic and social policies. Household mobility in response to economic incentives is critical, for example, to informed choice among regional growth and economic development policies, particularly for stagnant and declining regions. Similarly, the mobility response to economic incentives within urban areas affects both the efficiency and the distributional equity of such disparate activities as physical renewal and neighborhood preservation policies, intergovernmental fiscal relations, school finance, and local tax policy.

This chapter has two objectives. First, it attempts to isolate why a particular analytical problem—that of devising a structural model at the micro level of the mobility of households over space—has proved so intractable. It is argued that, in contrast to many superficially similar research problems, it is considerably more difficult to derive structural models with rejectable hypotheses about household mobility. In addition, the elements of a satisfactory model of short-distance (or intraurban) mobility are considerably more difficult to specify in theory and less amenable to measurement than are those of a model of long-distance (or interurban) mobility. Finally, the specific behavioral hypotheses arising from the theory of intraurban mobility are somewhat weaker than those arising from the theory of interurban mobility.

Second, despite the difficulty in specifying a fully satisfactory model of local mobility, this chapter argues that there are three distinct strands to mobility research which can illuminate policy choices at the local level, and that some simple steps could be taken to make mobility research more useful in the consideration of policy options at the local level.

MOBILITY OVER SPACE: THE ANALYTICAL PROBLEMS

At the most general level, households with given economic and demographic characteristics jointly choose a job, an employment location, a residence, a residential location, and a basket of "other goods," subject to a variety of financial, physical, and information constraints. Presumably, these choices are made in a rational, or at least systematic, way to maximize household well-being. At this level of abstraction, including a geographical dimension in the consumer choice problem yields no additional insight. It merely complicates an already elaborate description of self-interested behavior.

Describing the problem in its most general form suggests, however, one reason why the construction of micro models of geographic mobility is unusually demanding. The equilibrium solution to the general problem implies no residential or workplace mobility at all. Presumably, the flows of population between origins and destinations–whether scattered across regions or merely across city blocks–represent an equilibrating competitive force in the economy. In the static equilibrium world, with a given distribution of tastes among the population, the labor and housing markets across regions and within regions would equilibrate; the narrowing of price differentials would ensure that households would have no incentive to relocate in another region and that no household could be made better off by moving within any region.

Thus, theoretical and empirical analyses of mobility seek to represent an inherently dynamic or disequilibrium phenomenon. The dynamic nature of the mobility process is a source of difficulty in modeling relocation activity because little or no behavioral theory is explicitly dynamic. The typical mode of analysis is the comparison of static equilibrium states, especially by economists, or the comparison of stationary ergodic states, especially by geographers. Such comparisons suggest tendencies for movement "toward" a solution over "time." This methodology is, however, quite distinct from a truly dynamic analysis. The techniques of comparative statics, which constrast a set of initial and final conditions, can make no pretense of considering how those final conditions could be achieved. In this respect, the inherently dynamic nature of the phenomenon to be explained differs from the object of most other analyses undertaken by economists.

The general absence of theories of dynamic adjustment, not merely in economics but across the social sciences, is more of a penalty in the study of mobility—particularly the study of intraurban mobility—than in most other applications. The reason for this should be clear. For a large portion of the population, relocation in any reasonable time interval is a rather rare event (the average annual propensity to move is about .2, and repeat movers are a large portion of all movers). We should expect lags and inertia in responding to exogenous stimuli; in addition, the cost of gathering information about alternative locations is quite high, and the psychic and out-of-pocket costs of responding to these stimuli are large. Thus, the temporal relationship between a change in economic conditions and a subsequent mobility response is highly uncertain. We may expect long lags in adjustment. The determination of the magnitude and importance of these lags is a difficult empirical matter.

Describing the problem of mobility over space in its most general form—as the joint choice of a job and its location, a residence and its location, and a basket of other goods—also suggests the principal reason why empirical models of intraregional geographical mobility have proved more problematic than models of interregional mobility. Long-distance interurban or interregional moves typically involve decisions about job and employment choice. There is, therefore, some presumption that the decision to move and the choice of destination are motivated by investment considerations (Sjaasted, 1962). Thus, the analysis of long-distance mobility behavior generally proceeds by relating the place-to-place movement of individuals or groups to measures of the expected pecuniary costs of movement and the expected pecuniary benefits associated with different locations. Based upon market data on the costs of moving and the expected labor returns–information such as the earnings and employment probabilities of different classes of labor and the transport costs between different locations–predictive models of the mobility decision can be constructed and hypotheses can be tested directly.

In contrast to this theory and to the implied research strategy, short-distance intrametropolitan moves are quite frequently made without changes in the location or the type of job of the wage earner(s) in the household. Thus, the analysis of intrametropolitan mobility behavior generally proceeds by relating the propensity to move of individuals or groups to measures of the expected consumption benefits and costs, considering household income as exogenous. This consumption framework, which, it should be noted, neglects home ownership motives (which are partly investment motives) for intraurban moves, can be traced at least to Rossi (1955).

The conceptual and methodological distinction between long-distance and short-distance movement is important. By framing the long-distance move as an investment choice (at least to a first approximation), behavioral hypotheses can be tested in a more or less straightforward fashion using market information. Clearly, a model of mobility based upon this investment view is somewhat incomplete–surely there are consumption motives for choosing between, for example, San Francisco and New Haven, and surely these consumption tastes differ throughout the population (Weinberg, 1977). Nevertheless, by ignoring these motives entirely or by recognizing them in only a superficial way, models of interregional mobility have a remarkable predictive power and have proven useful in the consideration of a wide variety of policies, ranging from welfare reform to regional economic growth (Kain and Schaefer, 1972; Hansen, 1977).

In contrast, the consumption view of local mobility is much harder to translate into rejectable hypotheses about household behavior. To the extent that the economic theory of the household–utility maximization–is not tautological, it is because the arguments of consumers' preference functions (or at least the parameters of the demand curves for commodities) can be specified in a convincing way. However, consider a household with a given work place in an urban area. How does its basket of consumption differ if it relocates from one residential site to another? The consumption items that are specific to a residential site include the components of the particular dwelling units available across locations, the physical characteristics of the neighborhoods containing those different dwelling units, and the characteristics of the neighbors. In addition, the site-specific components of consumption include the quality and quantity of local public services supplied at different locations. Finally, the possibilities for consuming "other goods" (nonhousing, nonneighborhood goods) will also vary, to the extent that commuting costs at different sites leave more or fewer net resources available to the household.

The consumption costs of commuting at different locations can be derived directly from market observations. However, the consumption value of the components of the "residential services" available at different sites cannot be so derived. Thus, to represent the calculus of intraurban relocation in a reasonably satisfactory way, we require a detailed knowledge of the demand functions for the components of the residential services. (More specifically, knowledge of the demand functions permits estimates of the consumer surplus associated with consumption packages to be estimated.)

The demand functions indicate the precise relationships (for example, $Q = g[P, Y, A]$) among four kinds of variables: the amount of a commodity chosen by a household (Q), its price (P), household income (Y), and other relevant determinants of tastes (A). If we consider any component of the residential services consumed at a particular location (for example, the "quality of the neighborhood") it appears that only one of the four variables in the demand relation can be measured easily–household income. First, although everyone knows that there are "good" neighborhoods and "bad" neighborhoods and that households prefer better neighborhoods, there is little agreement about how to measure the quality of neighborhoods.

Second, although everyone knows that living in a better neighborhood costs more, the fact that neighborhoods are purchased jointly with all other residential services in a single transaction makes it difficult to infer the unit price of neighborhoods. Third, although everyone knows that the kind of housing preferred by households varies with household size and composition as well as income and price, there is little agreement on the specific factors to be included in an analysis, or to be "held constant" in the estimation of the demand relationship. Indeed, the use of the statistical jargon "held constant" by economists suggests that they have no particular expertise in defining the "other factors" affecting the market demand for components of the bundle of residential services.

This discussion is not to suggest that no one has attempted to estimate the demand curve for components of residential services (see, for example, Segal, 1979). It does suggest, however, that the estimation of these demands by any technique is based upon a number of strong assumptions or maintained hypotheses—stipulations about the definition and measurement of the components of housing services, about the form of jointness in the pricing of individual components of the rent bill, and about the "other factors" which affect demand. All of these considerations make measurement of a change in the preferred level of consumption of some component of residential services in response to a price increase far more problematic than measurement of a change in the preferred level of consumption of most other commodities traded in the economy.

In any case, estimates of the demand curves for the various components of residential services make it possible to estimate the difference between current and equilibrium consumption of each in response to exogenous changes in circumstances—for example, a change in household income or housing prices. With given moving costs, the probability of moving will vary monotonically with the level of disequilibrium for each component.

However, there is one further complication which makes the empirical analysis of intraurban mobility problematic. The heterogeneity of the bundle of residential services implies that information about alternatives is more costly to acquire for housing than for most other commodities. The heterogeneity of the commodity and the costliness of information suggest that there will be some price dispersion in the market even in the prices of "identical" dwelling units. This complicates the structural model of mobility: In response to an exogenous change in the desired or equilibrium configuration of residential services, a household may decide to search actively for an alternative dwelling unit. If, however, the household searches for a dwelling unit and the price of the unit sampled is high relative to the household's prior distribution of expected prices, the household may simply discontinue searching.

Understanding the mobility decision of individual households requires observations on the search and mobility choices made as well as on the socioeconomic characteristics of the households, the costs of relocating, and the residential services they consume (Weinberg, forthcoming). In studying individual choice, it is appropriate to view all these latter characteristics as given by market conditions. In studying the implications of mobility upon the economic geography of the urban area, this assumption is frequently inappropriate; indeed, the relationship between individual choice and collective outcome is often the essence of the problem. The general issue is "merely" a question of the measurement and extent of social externalities. Consider a neighborhood in equilibrium in the sense that each household has made the decision to live there on the basis of the average characteristics of the other residents as well as the characteristics of the dwelling units. Under these conditions, each resident household provides an externality to all other households in the neighborhood. What happens to the composition of the neighborhood if a few households move away? If the equilibrium is stable according to Schelling's (1978) definition, there may be no response to the perturbation. If the equilibrium is unstable, there may be a rapid change or turnover in neighborhood residents: A middle-income neighborhood may become a high-income or a poverty neighborhood as individual households move in response to the changed character of their neighbors—and the individual decisions of all who move change the character of the neighborhood for all others left behind.

LOCAL RESIDENTIAL MOBILITY: POLICY PROBLEMS

If we were to consider the types of actions undertaken by individual local governments in urban areas, we would find that virtually all have to do with the location and physical spatial arrangement of economic activity or with physical development as such. By this is included such diverse activities as the provision of urban transport (both public and private), the amount and location of investment in infrastructure, the level of provision of local services and their points of distribution, the promulgation of land use and zoning controls, the designation of recreational areas and public open spaces, and the location and extent of physical renewal and public housing. In addition, these decisions about the level, distribution, and location of public activities by individual jurisdictions imply differing types and levels of taxation–uniform with a jurisdiction, at least in principle, but varying widely among neighboring communities in the same metropolitan area. It is precisely these activities and public decisions which affect the costs and benefits of movement across urban neighborhoods in the same city, as well as the net benefits of moving from one jurisdiction to another.

It should be recognized that these are probably not the most important factors causing households to relocate in urban areas. Changes in family circumstances (such as income and household composition) may indeed be more important in affecting the probability of moving. Nevertheless, these public activities modify the character and quality of the residential services available at different locations or the prices at which such services are supplied. It is to be expected–indeed, in many cases it is intended–that local mobility responds to these actions of the local public sector. We can usefully distinguish three classes of information requirements for local decisions in which the insights of existing mobility research can be useful.

BASIC FACTUAL INFORMATION

For many public choices about these and other activities at the local level, the most pressing requirement is not for a sophisticated social science model of the causal factors underlying household inmigration, outmigration, and place-to-place movement. Indeed, for a great many policy considerations involving substantial allocations of public funds, the most important requirement is not even for a reduced form model capable of forecasting population flows a few periods into the future. Instead, the need is often for simple descriptive information—not why households move or even which households will move, but merely how many households have already moved. The constitutionally mandated census in the United States is taken each decade, and, during the long blackout that occurs in between, it is often difficult for local officials to know the characteristics of the populations which their policies are designed to serve.

A simple tabulation of the current population of a neighborhood or a jurisdiction in a single dimension—for example, age, income, number of workers, and so on—is frequently needed for current operations, not to mention for future planning purposes. For example, eligibility for many categorical grants (such as Community Development Block Grants) depends upon the size and demographic composition of resident populations. Indeed, in order to receive Community Development entitlements, civil divisions are required to estimate the current housing needs of lower-income households. There are literally dozens of similar instances in which choice among local policies depends merely on the current "facts" about local residents.

These facts need not be at issue. A unified population register would be sufficient to satisfy most of these questions of "fact," and the principal analytical questions would then be those of the information sciences—questions of efficient file organization, retrieval, and presentation of elements in the current register. However, population registers or annual population enumerations exist in only a few places outside Scandinavia, and even there they have not been used extensively (Ginsberg, 1973). With our Anglo-Saxon legal traditions, it is hardly conceivable that population registers, or even annual enumerations, would ever become widespread in North America.

Thus, a legitimate research challenge is the design of studies using collateral data merely to estimate the current population status of areas, neighborhoods, and jurisdictions. The general strategy includes at least three types of inquiry: first, the use of sample surveys to uncover current locational trends in a particular area and statistical methods to generalize conditions (Ericksen, 1973); second, the use of data sources on mobility from one location to make forecasts of the current mobility trends in other areas (Grier and Grier, 1977); and third, the use of iterative scaling and related techniques to estimate the elements of joint frequency distributions from marginal distributions (Apgar, 1976).

Despite the somewhat unglamorous nature of these forecasting procedures, there would be a great social, if not scientific, payoff to detailed comparisons between the actual changes in population observed from registers, enumerations, and decennial census data and those predicted from these classes of models designed to "forecast" current conditions.

SHORT-RUN FORECASTS

A second set of policy concerns is derived from the investment nature of so much of the activity of local government. Decisions about the amount and location of investments in urban infrastructure—transport systems, schools, hospitals, senior citizen centers, and so on—presuppose a set of locational comparisons over the economic lives of the facilities. These comparisons, in turn, are based upon projections or forecasts of the characteristics of the catchment population at alternative sites. The nature of particular investments, in turn, affects the attributes of catchment populations over time.

Estimates of these effects require some reasonably reliable reduced-form forecasting model of population change and urban relocation. Such reduced-form models may relate the characteristics of resident populations to one of several indices of mobility; for example, to estimates of the probability of moving from a local area, to the probability of moving to a local area, or, less frequently, to the probability of moving from one residential area to another. Such forecasting models relate the propensity to move to an array of population status variables, population change variables, and background characteristics of residential areas, using individual households as units of observation or using data averaged by neighborhood or census tract. By now there is an immense literature reporting the results of a large number of reduced-form statistical relationships between residential mobility and a loose collection of sociodemographic characteristics. Indeed, there are several extensive efforts to compare, contrast, and summarize various findings–those empirical regularities observed by different methods for different bodies of data at the individual level or at some aggregate level.

Some statements about the correlates of mobility can be made with a reasonable level of confidence, if not with a full understanding; certain demographic groups clearly have higher propensities to move than others. For example, age has a very high multiple partial correlation with residential mobility–both long-distance migration and short-distance relocation. Intraurban mobility is associated with changes in income and family size; the propensity to residential relocation varies over the household life cycle.

However, in the demography, sociology, geography, economics, and planning literatures there are at least four alternative definitions of the states comprising the household life cycle; and there are a large number of ways in which the complexities of household size, composition, age, and their changes could be plausibly represented. Despite this ambiguity, it appears that there is not a single study which attempts to replicate alternative reduced-form mobility models for different bodies of data, different locations, and/or different time periods. There are few areas in the social sciences where much attention is paid to replication. Most research is directed toward uncovering *the* (or yet another) cause, implying a different (or an additional) variable in a statistical model. It should therefore be stressed that regressions relating observations on moving or on place-to-place mobility to demographic characteristics are reduced-form relationships–a melange of supply and demand changes in the market for sites, complicated by inertia and lags in perception and information flows. Even with little a priori structure and only a loose behavioral theory, it would still be possible to exploit more fully the alternative hypotheses about the correlates of intraurban mobility by testing identical statistical relations across different sets of data.

STRUCTURAL RELATIONSHIPS

A third and more difficult set of policy concerns at the local level and at higher levels of government requires some explicit consideration of the structural model of residential mobility. The class of policy choices for which a recognition of the structure of mobility decisions is critical includes, for example, any substantial rearrangement of the distribution or method of finance of local public services.

The analysis of a substantial shift in the financing of services within the urban area (as, for example, with metropolitan-wide government and common tax rates adopted in Toronto and Jacksonville) or the analysis of a substantial shift in the levels of public service levels within the urban area (as threatened by school finance court suits pending under the equal protection clauses of various U.S. state constitutions) requires explicit consideration of the structure of local mobility.

In the former case, there may indeed be efficiency arguments for metropolitan consolidation—economies of scale in the planning and administration or in the production of government services. In the latter case, the objectives are more consciously redistributional—to increase the educational resources available to the children of poor households. In either case, however, the relationships among public service characteristics, taxes, and intraurban residential mobility suggest that the redistributional implications of reforms in financing or distributing local services are far more complex than was recognized initially by the proponents of reform.

The principal considerations, of course, are the incentives for mobility of such rearrangements in finance or service provision. The character of the local public services available at any site is an important component of the residential services consumed by choosing that site; the local tax is a part of the occupancy cost of that site. A rearrangement of services or their tax prices means that many households which had previously been content with their bundle of residential services, at existing prices, now find that this equilibrium has been disturbed exogenously. This disequilibrium may result in little residential relocation in the short run by households residing in the urban area.

Property values at different locations will, however, change more rapidly, if only because new migrants to the region will bid more for those locations where services have improved and taxes have declined than will the resident populations. In the aggregate, these changes in property values lead to changes in "grand lists," or the taxable value of real property, thus to changes in the tax rate choices of jurisdictions, and finally to changes in the level of public services which can be offered at a given tax price. Importantly, the mobility of households in response to these incentives has differential effects upon property values in various parts of the metropolitan area. Thus, even if such reforms are motivated by distributional concerns, and even if, on average, resources are transferred from high-income to low-income jurisdictions by such reforms, intraurban mobility may ensure that a large redistribution of wealth takes place between equally deserving households; that is, there is a great deal of capricious redistribution among homeowners and property owners of the same incomes and ex ante wealth.

As was argued in the introduction, a fully satisfactory model of household mobility and the demand for residential services is elusive; thus, empirical estimates of the long-run effects of such rearrangements and reforms are highly uncertain. Nevertheless, it is worth noting that in most of the early literature on school finance reform little or no attention was paid to the possible changes in residential locational patterns which might be induced by public action (Coons et al., 1970). Indeed, the court decision in *Serrano* v. *Priest* refers to variations in property values across districts (the outcome of the relocation process) as "fortuitous." Although the structural model of household mobility is indeed incomplete, these considerations have informed the legal (Friedman, 1977; Inman and Rubinfeld, 1979) and political debate about the desirability of various reforms (Bradford and Oates, 1974; McEachern, 1977).

Knowledge of the structural relationship between demands for residential services and local mobility is also critical to the design of policies seeking to affect neighborhood composition, revitalization, and the so-called "gentrification" or displacement phenomenon. Public concern with the latter issue, the displacement of poor households from their neighborhoods as a result of private renewal activity, has greatly increased over the past few years (Shur et al., 1977).

If the housing market is in equilibrium in a low-income neighborhood, this implies that, with their low incomes, resident households find it in their self-interest to purchase the smaller quantities of residential services available at that location. If the equilibrium of the neighborhood is disturbed—if, for example, higher-income households move in and rehabilitate some of the structures and if these activities produce external effects—then site values in the neighborhood may generally increase. To the extent that this happens, the quality of the neighborhood will have improved and the cost of residing there will naturally increase. Many low-income residents of the neighborhood, initially in equilibrium in their consumption of residential services, will find themselves consuming "too much" housing services. It is hard to conclude that low-income homeowners have been mistreated, since these exogenous changes in the neighborhood have increased the value of their homes and thus their wealth. For lower-income renter households, however, which have incentives to (are "forced to") move as neighborhoods are upgraded, it may be difficult to replicate their initial living conditions at similar prices. If for some reason we assume that some of these households (presumably older, poorer, long-term area residents) have something approaching a property right in their existing neighborhood, this calls for some form of subsidy to them to permit continued occupancy or to ease the transition to other residences. This argument based on mobility responses to neighborhood improvement may provide a rationale for transfers to some classes of renters, but it does not provide much insight about the appropriate amount.

THE LIMITS OF ANALYSIS

Consideration of formal models of residential mobility and the dynamics of neighborhood change suggest that there are sharp limits to the insights of social science. In most major cities, there is at least one neighborhood which has undergone dramatic physical improvement in the recent past. For a great many public planning purposes, it would be extremely useful to understand which areas or neighborhoods are "ripe" for such rapid physical upgrading. Such knowledge would clearly help in the design of physical renewal programs and the allocation of local funds across programs and neighborhoods. It would also help in estimating the number and character of households adversely affected by the renewal activity.

There apparently have been some attempts by scholars to develop planning models to predict the location and extent of this rehabilitation activity in urban areas (Sumka, 1978). It is difficult to imagine that these efforts could ever be successful in any realistic sense. Why? An entire "industry"—that of real estate speculation—is devoted to exactly this problem. There are enormous potential profits to anyone who could consistently isolate those blocks and neighborhoods ready to "take off," but profits among real estate speculators are not noticeably high. We should probably not expect such a highly specific and immediately practical understanding from scholarly research either.

CONCLUSION

Most of the research undertaken by social scientists is not related to public policy at all, or is related in only the most tangential way. Scientists and scholars are interested in an improved understanding of physical and social phenomena. Despite the fact that academic style, particularly among economists, dictates that each paper conclude with a statement about "policy implications," such pronouncements are seldom to be taken literally. Theoretical insights are based on *ceteris paribus* or *mutatis mutandis* assumptions, which make the application of findings to the real world typically quite indirect. Empirical estimates are rarely generalizable except as orders of magnitude. The distance between scholarly research and policy concerns does not make much difference to many researchers but is a source of frustration among some policy makers.

Research on the intrametropolitan mobility of households does not directly follow in this pattern. Of course, for many pressing policy questions—especially those facing local decision makers—social science research provides hardly a clue, while for other policy questions the solutions, in principle, are obvious and well known to researchers. There are some applications in which the models and statistical methods of mobility derived by social scientists can be used to improve policy formulation, or at least to make choices on the basis of better information. These applications involve estimates of the current socio-demographic characteristics of neighborhood or jurisdiction residents and short-run forecasts of demographic change. In addition, although a full understanding of the calculus of residential relocation is elusive, some qualitative insights into the effects of local policies can be made by considering the elements of a structural model. These insights are typically about the longer-run consequences of activity, after households have responded to their economic incentives by choosing sites for relocation. To make mobility research more useful than this, it seems, will require a far greater understanding of the interdependencies of household behavior in urban areas.

REFERENCES

APGAR, W. C., Jr. (1976) Housing Needs in New Hampshire. Report prepared for the New Hampshire Housing Commission and the New Hampshire Office of Comprehensive Planning.

BRADFORD, D. and W. OATES (1974) "Suburban exploitation of central cities and government structure," pp. 43-90 in H. M. Hochman and G. E. Peterson (eds.) Redistribution Through Public Choice. New York: Columbia University Press.

COONS, J. E., W. CLUNE, and S. SUGARMAN (1970) Private Wealth and Public Education. Cambridge, MA: Harvard University Press.

ERICKSEN, E. (1973) "A method of combining sample survey data and symptomatic indicators to obtain population estimates for local areas." Demography 10: 137-160.

FRIEDMAN, L. (1977) "The ambiguity of Serrano: two concepts of wealth neutrality." Hastings Constitutional Law Quarterly 4: 487-503.

GINSBERG, R. B. (1973) "Stochastic models of residential and geographical mobility for heterogeneous populations." Environment and Planning 5: 113-124.

GRIER, G. and E. GRIER (1977) "Using movership data to improve intercensal estimation of the population and housing market." Public Data Use 5: 11-19.

HANSEN, N. (1977) "Some research and policy implications of recent migration patterns in industrial countries." International Regional Science Review 2: 161-166.

INMAN, R. P. and D. L. RUBINFELD (1979) "The judicial pursuit of local fiscal equity." Harvard Law Review 92: 1662-1750.

KAIN, J. and R. SCHAEFER (1972) "Income maintenance, migration, and regional growth." Public Policy 20: 199–226.

MC EACHERN, W. (1977) School Finance Reform: The Implications of Tax Price Changes in Connecticut. Center for Real Estate and Urban Economic Studies. Real Estate Report 22. University of Connecticut.

ROSSI, P. (1955) Why Families Move. New York: Free Press.

SCHELLING, T. C. (1978) Micromotives and Macrobehavior. New York: W. W. Norton.

SEGAL, D. [ed.] (1979) The Economics of Neighborhoods. New York: Academic Press.

SHUR, R., C. HOLMAN, F. J. JAMES, et al. (1977) Hearings on Neighborhood Diversity. U.S. Senate, Committee on Banking, Housing and Urban Affairs. 7/7 and 7/8.

SJAASTED, L. (1962) "The costs and returns of human migration." Journal of Political Economy Supplement: 80-93.

SUMKA, H. J. (1978) Displacement in Revitalizing Neighborhoods: A Review and Research Strategy. Occasional Papers in Housing and Community Affairs Vol. 2. Washington, DC: U.S. Department of Housing and Urban Development.

WEINBERG, D. H. (1977) "Toward a simultaneous model of intraurban household mobility." Explorations in Economic Research 4: 579-592.

——— et al. (forthcoming) A Disequilibrium Model of Housing Search and Residential Mobility. Occasional Papers in Housing and Community Affairs. Washington, DC: U.S. Department of Housing and Urban Development.

Part II

Research Contributions to Understanding Mobility

☐ MOBILITY RESEARCH has concerned itself with both individual behavior and with the aggregate impact of residential moves on the composition of small areas within the city. Certainly much of this work has had little concern for public policy. Indeed, one of the consequences of its being such a pervasive phenomenon is that it has been absorbed into a variety of disciplinary perspectives on both individual and group behavior. The resulting fragmentation and largely unreconciled (although not necessarily irreconcilable) theoretical positions constitute a major barrier to translation of this work into public policy terms. The four chapters in this section reflect these contrasts in substantive focus and disciplinary orientation. They reflect an orthodox social science position which sees the role of the academic as informing policy by providing a better understanding of social process rather than by suggesting or reinforcing particular public actions. As a consequence, the links to public policy are somewhat more indirect and conjectural than in the subsequent sections of the volume.

All too often, theoretical discussions of residential location and relocation assume a stereotypical family unit with one wage earner to be the proper basis for analyzing decision-making behavior. As long as such households constitute the norm there are many insights to be obtained from such an assumption. However, its utility to public policy is limited to the extent that certain programs—particularly for the aged or for single-parent families—must be directed at "nonstandard" units.

Hughes' chapter forces us to recognize that recent demographic change has increased the divergence between theory and reality. Even since the last census, dramatic shifts have occurred in the characteristics of new household formation, in labor force participation, and in the growth of primary households. In light of the rapid inflation in housing prices and the costs of borrowing, we realize that changes are occurring in the societal context in which a wide variety of consumption and investment decisions, including mobility, are made. Not only do these changes possess important implications for public policy, particularly in terms of recognition of differential needs of social groups, but they demand to be incorporated in our theoretical arguments as well.

In Chapter 6 Michelson provides further support for the notion that it is not necessarily the norm which is relevant to public policy. Since a great deal of policy is driven by the public's perception of problems, it is appropriate to ask how mobility might be linked to current programmatic concerns. If we believe, as much of the literature suggests, that the bulk of residential mobility constitutes normal adjustment to changing household needs, then these actions define a context for policy but do not of themselves relate directly to problems tackled by policy. Michelson argues that we should pay more attention to those facets of mobility where the normal process of adjustments does not seem to be functioning—namely, where households are forced to relocate or where desired relocation is blocked by lack of resources or opportunity. His exploration of the nature of these circumstances and their implications remains consistent with conceptualizations of "normal" behavior within the sociological literature, yet is an innovative argument to link behavioral research to public policy.

Whatever differences arise in theoretical statements on the structure of individual movement decisions, there is no denying its complexity. The outcomes of decisions are a function not only of goals and preferences of households which vary as a function of life-cycle stage and accumulated housing experience but also of resources, opportunity, and information. When decisions are made under uncertainty and subject to a large number of contingent conditions, empirical research is not only difficult but very costly if analysis seeks to go beyond superficial relationships. One strategy for generating greater insights without incurring such expense is to simulate the decision process. The chapter by Smith and Clark adopts this path in an examination of the role of information in the search for and selection of a new home. Not only is this a difficult area in which to collect data and one therefore appropriate to simulation methods, but the substantive issue is of direct concern to those facing the problem of facilitating housing adjustments in heterogeneous markets in which information about opportunities is often restricted. It may transpire, for example, that expenditures on improved information about available low-income housing opportunities have a greater marginal effect on relocation outcomes than additional expenditures on housing allowances or other financial incentives.

The second major focus of mobility research is on the aggregate effect of residential moves on the composition of neighborhoods. Although we now appreciate that mobility is not the only mechanism of change, or even that change is a necessary correlate of mobility, it is still a major mechanism for neighborhood transition. Although we do not have the capability to predict change in detail except in special circumstances, such as those associated with large-scale urban renewal, the general processes of change in racial characteristics, income, and housing conditions and value are reasonably well understood. The work by the group at Washington University on St. Louis has been at the vanguard of research in this area, and Little's chapter here presents an important extension of that work. While a number of authors have expressed optimism that the rapid suburbanization of earlier decades is being reversed with new flows of the more affluent back into the city, Little demonstrates that, at least in St. Louis, the dominant dynamic does not seem to have changed. His work suggests that the now-familiar arbitrage cycle which saw the gradual transformation of the inner city is now being repeated in the suburbs. The driving force might still be seen as the large commitment of senior levels of government to the continuation of the development process and its associated payoffs to consumption-oriented industry, rather than the attempt to promote reinvestment in the inner city.

5

Demographic and Economic Change: Implications for Mobility

JAMES W. HUGHES

☐ AS THE 1970s DREW TO A CLOSE, the nation was beset by a number of policy issues which originated or were discerned during the decade. While energy, environment, regional shifts, and rising housing cost thresholds have secured prominent notice, an equally compelling set of interrelated events has yet to receive ample scrutiny: the mobility implications of the processes of household evolution, labor force participation, income growth patterns, and shifts in housing tenure (owner versus rental). While the trendlines have been separately analyzed, there have been only tentative steps taken toward logical integration and synthesis. This is certainly understandable, however, given available data sources which, while amply documenting each phenomenon individually, do not readily facilitate the specification of their interrelationships.

The general objective of this chapter is to explore the possible ramifications for residential mobility of the confluence of these events. As inflation and income lags appear to represent endemic conditions for the immediate future, there may well be substantial impediments raised to certain types of residential mobility, restrictions which would have major impacts on the current social fabric. More specifically, I will focus attention on two types of mobility: the rate of household formation and shifts in housing tenure within the framework of labor force participation and income changes. By definition, household formation is a form of mobility, since it involves establishing residence within a separate dwelling unit. Similarly, tenure shifts involve moving between rented and owned accommodations, again, in most cases representing a change of dwelling units. Although new household formation and shifts in tenure may not account for more than 40 percent of local moves, they possess a great deal of social significance.

As Norman and Susan Fainstein point out later in this volume, mobility has traditionally been regarded as the American way of improving one's circumstances. Nowhere is this more important than at the times when a separate residence is established for the first time or when the first move is made to acquire one's own house. The basic thesis of this study is that we already have evidence to suggest that important changes are taking place in the social, demographic, and economic structure of households, changes which are transmitted via the processes of mobility to produce significant social effects in the allocation of housing. These changes tend to be omitted from most discussions of the dynamics of urban housing markets; the evidence presented here argues strongly for their inclusion.

ORGANIZATION

Initially, I will examine the broad patterns of household change, focusing on the shifting compositional profile as well as changing size parameters. Concurrent developments in labor force participation are then considered, with the data partitioned by marital status. Variations in income across the profile of household types are subsequently presented, with attention directed to the sharp differentials which emerge. Finally, the relationships between household configurations and housing attributes are studied, with the analysis centering on tenure shifts, income and income-cost variations.

HOUSEHOLD EVOLUTION

The population of the United States is in the midst of a long-term process of rapid segmentation into an increasing number of varied household types. Indeed, the pace of the phenomenon has been so intense as to give the appearance of a slow-growing population immediately diffusing itself into any available expansion of the housing supply. For example, while the population increase of the 1970s probably will be 25 percent less than that registered during the 1960s, the net housing additions should be about 60 percent greater.[1] The process is marked not only by decreasing household sizes, but also by the increasing significance of what were once considered "atypical" configurations. The basic outlines of the latter shifts are contained in the data of Table 5.1.

Family households (two or more related individuals sharing a dwelling unit) accounted for 85 percent of all households in 1960; by 1978, their proportional share declined to 74.9 percent. Within this broad category, however, are disparate growth patterns. Husband-wife families, the once classic modal form, had the slowest growth rate (5.9 percent) over the 1970-1978 period. By 1978, they represented 62.3 percent of all households, a substantial reduction from their 74.3 percent share of 1960. Expanding at much faster rates are male and female family householders (no spouses present), although their absolute change (combined) is of similar magnitude to husband-wife families over the 1970-1978 period.

The most dramatic growth is exhibited by nonfamily households, which increased their representation within the profile of household types from 15.0 percent in 1960 to 25.1 percent in 1978. Much of the absolute gain was by single-person households (householders living alone); but households with nonrelative(s) present (households comprising two or more unrelated individuals) showed the greatest percentage change (115.4 percent) between 1970 and 1978.

An equally telling indicator is household size, whose current changes represent a corollary to the process of household evolution depicted above. The data in Table 5.2 have been partitioned by housing tenure with both owner-occupied and rental sectors reflecting a considerable reduction in household size over the brief period from 1970 to 1976, and both are indicative of the longer-term trend.[2] In the rental sector, only the one- (34.0 percent) and two- (17.1 percent) person categories reveal any appreciable growth, while the larger arrangements

TABLE 5.1 Households by Type and Size: 1978, 1970, 1960
(Numbers in thousands, noninstitutional population.)

Subject	1978		1970		1960		Percent Change	
	Number	*Percent*	*Number*	*Percent*	*Number*	*Percent*	*1970 to 1978*	*1960 to 1970*
HOUSEHOLDS								
Total households	76,030	100.0	63,401	100.0	52,799	100.0	19.9	20.1
Family households	56,958	74.9	51,456	81.2	44,905	85.0	10.7	14.6
Husband-wife	47,357	62.3	44,728	70.5	39,254	74.3	5.9	13.9
Male householder, no wife present	1,564	2.1	1,228	1.9	1,228	2.3	27.4	–
Female householder, no husband present	8,037	10.6	5,500	8.7	4,422	8.4	46.1	24.4
Nonfamily households	19,071	25.1	11,945	18.8	7,895	15.0	59.7	51.3
Householder living alone[a]	16,715	22.0	10,851	17.1	6,896	13.1	54.0	57.4
Householder with nonrelative(s) present	2,356	3.1	1,094	1.7	999	1.9	115.4	9.5

[a]One-person household.

SOURCE: U.S. Bureau of the Census, *Current Population Reports,* Series P-20, Nos. 324 and 327, and unpublished Current Population Survey data.

TABLE 5.2 Persons Per Household (Household Size): 1970-1976
United States Total (Numbers in thousands)

	1970		*1976*		*Change: 1970 to 1976*	
	Number	*Percent*	*Number*	*Percent*	*Number*	*Percent*
OWNER-OCCUPIED						
Total	39,886	100.0	47,904	100.0	8,018	20.1
1 person	4,762	11.9	6,278	13.1	1,516	31.8
2 persons	12,010	30.1	15,098	31.5	3,088	25.7
3 persons	6,985	17.5	8,677	18.1	1,692	24.2
4 persons	6,925	17.4	8,786	18.3	1,861	26.9
5 persons	4,554	11.4	5,075	10.6	521	11.4
6 persons	2,468	6.2	2,341	4.9	–127	– 5.1
7 persons or more	2,182	5.5	1,649	3.4	–533	–24.4
Median	3.0	–	2.8	–	–	–
RENTER-OCCUPIED						
Total	23,560	100.0	26,101	100.0	2,541	10.8
1 person	6,389	27.1	8,560	32.8	2,171	34.0
2 persons	6,773	28.7	7,929	30.4	1,156	17.1
3 persons	3,923	16.7	4,036	15.5	113	2.9
4 persons	2,875	12.2	2,846	10.9	– 29	– 0.1
5 persons	1,643	7.0	1,443	5.5	–200	–12.2
6 persons	915	3.9	688	2.6	–227	–24.8
7 persons or more	1,043	4.4	599	2.3	–444	–42.6
Median	2.3	–	2.1	–	–	–

Notes: **Numbers and/or percentages may not total 100 due to rounding.**
SOURCE: U.S. Department of Commerce, U.S. Bureau of the Census (1978).

(six persons or more) are rapidly contracting in number. Indeed, the scale of the attrition among very large households was one of the more dramatic social changes occurring in the 1970s.

As we proceed into the decade of the 1980s, then, there appears to be considerable momentum to the basic transformation of the American household, with what were once considered "atypical" households securing an increasing presence and with visible reductions in scale attendant on all household sectors. The end result, the rapid formation of households, is in itself a form of mobility, underlaid by the following factors:[3]

(1) a long-term decline in the fertility rate and an increasing rate of childlessness among young married women;
(2) declining birth expectations of women, particularly among the younger age cohorts;
(3) fundamental changes in marriage relationship–rising divorce rates, declining marriage rates, and postponement of marriage by young women; and
(4) changing labor force patterns acting in concert with the preceding elements.

While this list is certainly an abbreviation of the factors at work, the last-named phenomenon, labor force participation, deserves further consideration.

LABOR FORCE SHIFTS

The escalating pace of independent household formation probably cannot occur in the absence of increases in the economic wherewithal of the actors involved in the process. The labor force shifts detailed in Table 5.3 give some indication of the correspondence between these two variables. Most striking are the absolute labor force gains and increasing participation rates among women, particularly among the categories of "never married" and "married, husband present." The former is at the forefront of the surge in nonfamily households, while the latter relates to the economic capacity of traditional husband-wife families. Also of significance are the labor force increases evidenced by divorced women, which correspond to the rise of female-headed families and/or households (female householders, no husband present) and is linked to the process of family fragmentation.[4]

Among men, the "married, wife present" category exhibits a sharp decline in the rate of participation, reflecting the increasing prevalence of early retirements. The most significant gains are attached to those "never married" (linked to nonfamily households) along with the "married, wife absent" and "divorced" sectors, which correspond to both nonfamily households and families with no wife present.

TABLE 5.3 Civilian Labor Force by Sex and Marital Status, March 1970 and 1978

	Civilian Labor Force			
	Number (in thousands)		*Labor Force Participation Rate*	
Sex and Marital Status	*March 1970*	*March 1978*	*March 1970*	*March 1978*
Both sexes, total	81,693	98,437	59.1	62.2
Men, total	50,460	57,466	77.6	76.8
Never married	9,421	13,978	60.4	69.2
Married, wife present	38,123	38,507	86.6	81.6
Married, wife absent	1,053	1,703	61.3	77.4
Widowed	672	657	31.9	30.5
Divorced	1,191	2,711	76.0	80.7
Women, total	31,233	40,971	42.6	49.1
Never married	6,965	10,222	53.0	60.5
Married, husband present	18,377	22,789	40.8	47.6
Married, husband absent	1,422	1,802	52.1	56.8
Widowed	2,542	2,269	26.4	22.4
Divorced	1,927	3,888	71.5	74.0

Note: Because of rounding, sums of individual items may not equal totals.
SOURCE: Johnson (1979).

While household shifts and labor force changes show considerable correspondence—at least as revealed by the surface data presented here—we can only speculate as to the causes. Even then, the directional flows remain problematic; for example, do the labor force participation patterns of women facilitate divorce and household formation, or do the latter necessitate labor force participation?

A more detailed examination of women in the labor force in 1978 can be made using Table 5.4, which partitions the parameters of Table 5.3 by the presence of children and their ages. While more extensive analysis is possible, it is sufficient to note here that the highest participation rates are exhibited by those women with children six to 17 years of age across all marital statuses. Rather than being an impediment to women working, children appear to represent more of an economic requirement to labor force participation. This reflects two important considerations relevant to this analysis: the economic constraints of

TABLE 5.4 'Labor Force Status of Women 16 Years and Over, by Marital Status and Presence and Age of Youngest Child: March 1978 (numbers in thousands)

Marital and Labor Force Status	Total	Number Children Under 18 Years	With Children Under 18: Total	6 to 17 Years Only	Under 6 Years
Women, 16 and over, total	83,374	52,861	30,513	17,213	13,299
in labor force	40,971	24,823	16,147	10,334	5,813
Labor force participation rate	49.1	47.0	52.9	60.0	43.7
Never-married total	16,891	15,951	940	303	637
in labor force	10,222	9,746	476	212	264
Labor force participation rate	60.5	61.1	50.6	69.9	41.5
Married, husband present	47,906	23,066	24,841	13,694	11,147
in labor force	22,789	10,320	12,469	7,829	4,640
Labor force participation rate	47.6	44.7	50.2	57.2	41.6
Married, husband absent	3,173	1,495	1,679	913	765
in labor force	1,802	818	985	564	420
Labor force participation rate	56.8	54.7	58.6	61.8	54.9
Widowed, total	10,147	9,470	677	595	82
in labor force	2,269	1,891	379	337	41
Labor force participation rate	22.4	20.0	55.9	56.7	50.3
Divorced, total	5,257	2,880	2,377	1,709	668
in labor force	3,888	2,049	1,839	1,391	447
Labor force participation rate	74.0	71.2	77.4	81.4	67.0

Note: Because of rounding, sums of individual items may not equal totals.
SOURCE: Johnson (1979).

independent household maintenance and the felt need of two-worker families (or households) to secure the attributes of the "good life," particularly homeownership.

The latter would appear to add complexity to residential locational decision-making, particularly when the need for two workers coincides with child-rearing responsibilities. Choices are not only bound by stage in family lifestyle considerations, such as the desire for good schools, but also by labor market constraints which affect the availability of employment opportunities attractive to both spouses.[5]

At the same time, the rise of "atypical" households, in conjunction with expanding labor force participation, would appear to portend residential arrangements and choices free from the constraints of childbearing criteria. However, such mobility may fail to materialize in an era of sustained inflation, particularly if incomes lag.

INCOME VARIATIONS

The economic realities of the intersection of household changes and labor force shifts are revealed by patterns of income resources. The broader trends in this regard are detailed in Table 5.5. Families, despite their sluggish growth in absolute numbers, reside at the upper end of the income spectrum. Their median 1977 income, $16,009, represents a $6142 increment above the 1970 base. The income of unrelated individuals (which include nonfamily households as well as individuals residing in group quarters) continues to lag markedly, despite the higher rate of growth over the 1970-1977 period.

All households (which include family and nonfamily households, but exclude primary individuals residing in group quarters) also lag behind the family income parameters as to absolute level, absolute growth, and rate of change. Hence, while the social functions of families may be diminishing in an era of low fertility levels, it is difficult to deny their economic role. In an inflationary milieu when real income growth in constant dollars is minimal, families may well be able to cope more effectively than other household types, at least in the aggregate.

However, sharp income differentials exist within the array of family types, as shown by the data in Table 5.6. By all measures, husband-wife families exceed the income resources of their male- and female-headed counterparts (spouse absent).[6] Further, within husband-wife configurations, families in which the wife is in the paid labor force have the highest income levels. In 1977, such families had a median income of $20,268 as compared with $16,009 for all families and $15,063 for husband-wife families (with wife not in paid labor force). This

TABLE 5.5 Income Changes: Families, Unrelated Individuals, and Households (U.S. Total), 1970-1977

Unit	*1970*	*1977*	*Change: 1970 to 1977* *Number*	*Percent*
Families	$9,867	$16,009	$6,142	62.2%
Unrelated individuals	3,137	5,907	2,770	88.3
Families and unrelated individuals	8,335	12,666	4,331	52.0
Households	8,734	13,572	4,838	55.3

SOURCE: U.S. Bureau of the Census (1971, 1979).

TABLE 5.6 Income Shifts by Family Type, U.S. Total 1970-1977

Family Type	*1970*	*1977*	*Change: 1970 to 1977* *Number*	*Percent*
Total Families	$ 9,867	$16,009	$6,142	62.2%
Male head total	10,480	17,517	7,037	67.1
Married, wife present	10,516	17,616	7,100	67.5
Wife in paid labor force	12,276	20,268	7,992	65.1
Wife not in paid labor force	9,304	15,063	5,759	61.9
Other marital status	9,012	14,518	5,506	61.1
Female head	5,093	7,765	2,672	52.5

SOURCE: U.S. Bureau of the Census (1971, 1979).

wide range of income variation may well exemplify a new class partition within the United States: households with one versus two or more persons working.[7]

It is useful to recast these family configurations in terms of their absolute order of incidence (Table 5.7). As previously mentioned, husband-wife families represented the slowest-growing household configuration. However, this overall category masks substantial internal variation. The subset with wives in the paid labor force grew by 24.9 percent (4.4 million husband-wife families), while there was an absolute decline (-6.3 percent) in husband-wife families with wives not in the paid labor force. Thus, the family partition with the greatest increase in absolute numbers is the most income-affluent subset.

TABLE 5.7 Family Configurations, U. S. Total 1971-1978
(numbers in thousands)

Family Type	*March 1971*		*March 1978*		*Change: 1971 to 1978*	
	Number	*Percent*	*Number*	*Percent*	*Number*	*Percent*
Total Families	51,948	100.0%	57,215	100.0%	5,267	10.1%
Male head total	45,998	88.5	48,979	85.6	2,981	6.5
Married, wife present	44,739	86.1	47,385	82.8	2,646	5.9
Wife in paid labor force	17,568	33.8	21,936	38.3	4,368	24.9
Wife not in paid labor force	27,172	52.3	25,449	44.4	–1,723	– 6.3
Other marital status	1,258	2.4	1,594	2.8	336	26.7
Female head	5,950	11.5	8,236	14.4	2,286	38.4

SOURCE: U.S. Bureau of the Census (1971, 1979).

Of equal significance are female-headed families, which demonstrated the greatest rate of increase (38.4 percent) and the second highest absolute gain (2.3 million families). Consequently, a sharp income bifurcation is forming within the family categorization, with the most income-deficient configuration (female-headed families) also representing a sector of dramatic growth. Further partitioning of the data reveals even more disturbing variations. Most onerous is the position of black female-headed families residing in central cities, whose 1977 median income was only $5125, far below that of all female-headed families ($7765). With such abysmal income resources, the capacity for voluntary mobility is virtually nonexistent.[8]

HOUSING TENURE AND RISING COSTS OF OWNERSHIP

The disparity in income levels across the various household formats is closely associated with shifts in housing tenure documented over the period 1970 to 1976. While the data in Tables 5.8 and 5.9 reveal variations in market penetration by household type for renters and owners separately, the correlation of household type with income allows some inference to be made regarding earning capacity and this particular form of mobility.

From Table 5.8 it is apparent that the rental sector is losing the most affluent household group—namely, husband-wife families (–11.5 percent).[9] In contrast, the owner-occupied sector

(Table 5.9) added 5.3 million (17.2 percent) such families over the six-year period under consideration. Other male-headed households had a rate of increase in rental facilities double that of the owner-occupied sector.[10] Similarly, female-headed groups had a far higher rate of growth in rental units than held true in owner-occupied accommodations. Thus, those households characterized by more substantial income resources are increasingly attracted to homeownership; rental facilities are increasingly the refuge of household configurations of lower income.[11] The rental market is being "cream skimmed" of its more affluent tenantry.

The decision-making rationales underlying this pattern of tenure shift, which has, if anything, intensified since 1976, have by now assumed the status of an established convention, if not folk wisdom. Purchasing a house is no longer based primarily on acquisition of shelter but rather on acquiring an investment or a form of tax shelter as one of the few successful refuges from an inflationary environment. We have become a postshelter society. Increasingly excluded from this society, and hence from the associated mobility consequences, are those lacking the economic means to secure membership.

The inflation in housing costs for homeowners is essentially counterbalanced by a parallel inflation in real or conjectured capital values. In terms of immediate cash flow demands, homeowners—particularly newcomers to the market—seemingly are willing to devote unprecedented portions of their income to supporting the purchase of housing. The advantages are nominally tax relief and potential for capital gains. Thus, when homeownership is employed as a leveraged investment, its support can be accepted quite realistically as the sum of nominal shelter costs plus the surrogate amount of funds which would once have gone into more traditional forms of investments: private insurance, savings accounts, and the securities market.

The rental housing situation is quite different: There are no tax advantages to the tenant; rents must be paid from posttax income. Moreover, there is no compensatory investment element offsetting rents. The rental market and its soaring cost structures, both in terms of development and operation, must

TABLE 5.8 Renter-Occupied Units Household Composition by Age of Head 1970-1976 (numbers in thousands)

Household Composition by Age of Head	*1970*	*1976*	*Change: 1970 to 1976*	
			Number	*Percent*
Total renter-occupied	23,560	26,101	2,541	10.8
2 or More Person Households	17,171	17,541	370	2.2
Male head, wife present	12,759	11,291	–1,468	–11.5
Under 25 years	2,282	2,210	– 72	– 3.2
25 to 29 years	2,408	2,440	32	1.3
30 to 34 years	1,531	1,459	– 72	– 4.7
35 to 44 years	2,154	1,722	– 432	–20.1
45 to 64 years	3,148	2,332	– 816	–25.9
65 years and over	1,236	1,127	– 109	– 8.8
Other male head	1,143	1,730	587	51.4
Under 65 years	1,010	1,642	632	62.6
65 years and over	132	88	– 44	–33.3
Female head	3,270	4,520	1,250	38.2
Under 65 years	2,899	4,130	1,231	42.5
65 years and over	370	391	21	5.7
1-Person Households	6,389	8,560	2,171	34.0
Under 65 years	4,109	5,803	1,694	41.2
65 years and over	2,279	2,757	478	21.0

Note: Numbers and/or percentages may not add due to rounding.
SOURCE: U.S. Department of Commerce, U.S. Bureau of the Census (1978).

compete within an environment devoid of the incentives attached to homeownership.

The result of this process by 1976 is partially revealed in Table 5.10, which compares the incomes of corresponding household categories residing in owner- and renter-occupied facilities. The income differentials are apparent and pervasive; those who occupy rental facilities have far lower incomes. Furthermore, the renter household type, which is close in income to its owner counterpart, the husband-wife family, is the one which is apparently vacating the rental market at the greatest rate.

TABLE 5.9 Owner-Occupied Units Household Composition by Age of Head 1970-1976 (numbers in thousands)

Household Composition by Age of Head	*1970*	*1976*	*Number*	*Percent*
Total owner-occupied	39,886	47,904	8,018	20.1%
2 or More Person Households	35,124	41,626	6,502	18.5
Male head, wife present	30,806	36,108	5,302	17.2
Under 25 years	800	1,069	269	33.6
25 to 29 years	2,252	3,186	934	41.5
30 to 34 years	2,938	4,069	1,131	38.5
35 to 44 years	7,097	7,512	415	5.8
45 to 64 years	13,230	14,743	1,513	11.4
65 years and over	4,490	5,530	1,040	23.2
Other male head	1,298	1,629	331	25.5
Under 65 years	974	1,251	277	28.4
65 years and over	324	378	54	16.7
Female head	3,019	3,889	870	28.8
Under 65 years	2,159	2,922	763	35.3
65 years and over	860	967	107	12.4
1-Person Households	4,762	6,278	1,516	31.8
Under 65 years	2,075	2,677	602	29.0
65 years and over	2,688	3,602	914	34.0

Note: Numbers and/or percentages may not add due to rounding.
SOURCE: U.S. Department of Commerce, U.S. Bureau of the Census (1978).

Further amplifying the economic disparities are the income-rent (Table 5.11) and income-house value (Table 5.12) relationships disaggregated by household type. Within the rental sector (Table 5.11), it is the husband-wife category which must surrender the smallest proportion (18.9 percent) of its income for rent. Female-headed households (two or more persons), in contrast, exhibit the most stress, particularly those under 35 years of age, with the median annual rent consuming over 35 percent of median annual income.

Within the owner-occupied sector, the husband-wife household again commands the most advantageous economic position, possessing the lowest house-value-to-income ratio while residing in the highest value units.[12] For male-headed and one-person households, the reverse relationship holds: They

TABLE 5.10 Owner- and Renter-Occupied Household Income, by Household Configuration: U.S. Total, 1976

	Owner-Occupied	*Renter-Occupied*	*Renter/Owner Ratio*
Total Households	$14,400	$ 8,100	.56
2 or More Person Households	15,900	9,500	.60
Male head, wife present	16,800	11,600	.69
Under 25 years	12,800	9,800	.77
25 to 29 years	16,200	12,600	.78
30 to 34 years	17,900	13,000	.73
35 to 44 years	19,700	13,200	.67
45 to 64 years	18,600	12,600	.68
65 years and over	8,400	6,600	.79
Other male head	14,300	8,500	.59
Under 65 years	15,800	8,600	.54
65 years and over	9,000	6,000	.67
Female head	9,100	5,600	.62
Under 65 years	9,600	5,700	.59
65 years and over	7,100	5,100	.72
1-Person Households	5,100	5,500	1.08
Under 65 years	8,600	7,800	.91
65 years and over	4,200	3,400	.81

Note: Numbers and/or percentages may not add due to rounding.
SOURCE: U.S. Department of Commerce, U.S. Bureau of the Census (1978).

reside in the lowest-value houses but have the highest house-value-to-income ratio.

The attenuation of America's population into an even greater number of smaller households suggests that income resources are being spread too thin over an expanded housing supply. With housing costs escalating much more rapidly than income, despite unparalleled growth in the labor force, this view may gain increasing validity.[13] If two or more worker households become the threshold condition for sustaining house-buying power, then we may be reaching the limits to the phenomenon of unimpeded household formation and the loss of mobility such limits imply.

TABLE 5.11 Renter-Occupied Units, by Household Configuration: U.S. Total, Income and Rent Relationship, 1976

Household Configuration[a]	*1976 Median Income*	*1976 Monthly Median Gross Rent*	*1976 Annual Median Gross Rent*	*1976 Median Rent as a Percent of Median Income*
2 or More Person Households	$ 9,500	$179	$2,148	22.6%
Male head, wife present	11,600	183	2,196	18.9
Under 25 years	9,800	166	1,992	20.3
25 to 29 years	12,600	192	2,304	18.3
30 to 34 years	13,000	193	2,316	17.8
35 to 44 years	13,200	195	2,340	17.7
45 to 64 years	12,600	184	2,208	17.5
65 years and over	6,600	163	1,956	29.6
Other male head	8,500	186	2,232	26.3
Under 65 years	8,600	188	2,256	26.2
65 years and over	6,000	144	1,728	28.8
Female head	5,600	167	2,004	35.8
Under 65 years	5,700	168	2,016	35.4
65 years and over	5,100	141	1,692	33.2
1-Person Households	5,500	142	1,704	31.0
Under 65 years	7,800	155	1,860	23.8
65 years and over	3,400	108	1,296	38.1

[a]No nonrelatives present.
SOURCE: U.S. Department of Commerce, U.S. Bureau of the Census (1978).

SUMMARY

The momentum of these tendencies, if sustained, can lead to an ominous future scenario, one which possesses serious ramifications for mobility. If we assume the high inflation rates of the past decade are here to stay, and if the household configurational transformation does not suffer any sharp alterations, then the income differentials will become more deeply etched into the nation's housing markets. Those households which have not done so will vacate the rental market if their economic means permit, entering into the ranks of homeowners. This will further deplete the market pool of rental housing of its more income-sufficient tenants.

TABLE 5.12 Owner-Occupied Units, by Household Configuration: U.S. Total, Income and House Value Relationship, 1976

Household Configuration[a]	*1976 Median Income*	*1976 House Value*	*House Value-Income Ratio*
2 or More Person Households	$15,900	$33,400	2.1
Male head, wife present	16,800	34,300	2.0
Under 25 years	12,800	25,400	2.0
25 to 29 years	16,200	31,800	2.0
30 to 34 years	17,900	37,000	2.1
35 to 44 years	19,700	38,000	1.9
45 to 64 years	18,600	34,900	1.9
65 years and over	8,400	28,100	3.3
Other male head	14,300	30,000	2.1
Under 65 years	15,800	31,900	2.0
65 years and over	9,000	23,100	2.6
Female head	9,100	25,900	2.8
Under 65 years	9,600	26,600	2.8
65 years and over	7,100	23,300	3.3
1 Person Households	5,100	23,500	4.6
Under 65 years	8,600	25,700	3.0
65 years and over	4,200	21,900	5.2

[a]No nonrelatives present
SOURCE: U.S. Department of Commerce, U.S. Bureau of the Census (1978).

The maturation of a "postshelter society" in which housing is viewed not so much as shelter from the elements but as a refuge from inflation would appear to affect mobility in several ways. Private rental housing will become an increasingly scarce commodity. A pool of tenants lacking the income necessary to meet the cost thresholds attendant on new construction, in conjunction with the imminent appearance of the baby bust generation (a decreasing number of new household formations), severely reduces the potential and market rationale of significant new private rental construction. In addition, the existing supply may tend to contract if condominium conversions continue to increase as a strategy for capturing at least part of the withdrawal of more affluent tenantry.

The end effect would prove to be incentives to stability in place, with the search for new rental accommodations–in present or different spatial locales—an increasingly difficult task. Furthermore, long-term tenants will find their ownership aspirations increasingly thwarted by a housing price structure escalating in concert with, or above, the inflation rate, thus constraining tenure-related mobility.

Households fortunate to be owners may also find restraints to mobility. The pattern of multiple workers necessary to support an "appropriate" home serves as a constraint in itself, with destinations limited to those labor markets with sufficient opportunities attractive to the working members. This case would arise most strongly in situations where both spouses are on independent career "tracks."

While the latter grouping may possess the economic capacity for mobility, both in terms of income flows and built-up equity as a result of ownership, the mortgage market may offer additional impediments. Whether equity accumulation permits either stability or upgrading in terms of the quantity and quality of housing consumed, it becomes difficult to accept higher-rate financing, which in most cases is an unavoidable by-product of mobility. In fact, a further complication is the possibility of being "stuck" in a soft market, where inflation fails to act as a "take out" mechanism—that is, when resale prices become insufficient to recover that capital required to enter a new geographic area. Thus, even for the more affluent class of households (those owning homes and having two persons in the labor force) constraints to mobility may also increase. Of course, if inflation keeps advancing at the current rate, the financial aspect of these constraints will last only until such time as one income becomes sufficient to maintain the housing payments incurred at the beginning of the mortgage period. One might also expect innovations in the mortgage instrument incorporating escalating, inflation-dependent schedules to allow the financial burden to be more easily absorbed.

The major point to arise from this discussion is not that we can predict new and significant forms of mobility, but that there are rapid and substantial changes occurring in the social, demographic, and economic fabric of society, changes which are having and will continue to have impact on the dynamics of urban housing markets. As yet, they lie outside the conventional framework of analysis for these markets. Yet, as has been shown here, they are so strong that they cannot continue to be ignored.

NOTES

1. From 1960 to 1970, the nation's population increased by approximately 24 million people. Between 1970 and 1980, the net increase will be in the vicinity of 18 million people. In contrast, the net gain of housing units over the 1960-1970 period totaled approximately 10 million units; between 1970 and 1980, the increase should total approximately 16 million units. The decade of the 1970s appears to have been the most prolific housing production period in the nation's history.

2. The average household size in 1950 was 3.37 persons. There has been an uninterrupted decline over the ensuing 28 years. By 1978, the average household size had declined to 2.81 persons.

3. Again, it is interesting to note the changing differential between population and household growth rates. From 1960 to 1970, the nation's population increased by 13.3 percent; the number of households increased by 20.1 percent. Between 1970 and 1978, the rate of population growth approached only 6.8 percent. However, in this eight-year period, the growth in the number of households (19.9 percent) nearly matched the rate of the entire preceding decade. Certainly, the availability of housing facilitated the apparently unimpeded expansion of the nation's household ranks.

4. The married (spouse absent), widowed, and divorced statuses can be linked to both households and families, depending upon whether any additional household members are relatives (families) or nonrelatives (households).

5. Thompson (1978) has suggested area revitalization strategies centering on labor force policies which would maximize the employment opportunities for educated women, thus helping to attract two-worker families.

6. Recent Census Bureau policy plans to discontinue the use of the term "head" in favor of "householder." The terminology in this chapter will reflect that presented in the data tabulations, which in certain instances either predate present policy or have yet to be changed due to technical reasons.

7. The reason for the income differential not being even wider is because only 17 percent of all husband-wife families had both spouses working year-round full-time in 1977 (Rawlings, 1978).

8. For a detailed analysis, see Sternlieb and Hughes (1979).

9. Obviously, the data mask the underlying dynamics and do not reveal the direct transitions from one tenure status to the other, the formation of new households, and the demise of others. However, the magnitude of the shifts incurred make it reasonably safe to infer that these households endowed with higher incomes are shifting to owner occupancy.

10. Unfortunately, ambiguity arises with the Annual Housing Survey classifications (as compared with the Current Population Survey categories). Male- or female-headed (no spouse indicated), two or more persons households can be either families or households, depending on whether or not the other household members are relatives (families) or nonrelatives (households).

11. The one-person ratio as shown is correct; it probably can be viewed as a result of the sensitive nature of medians resulting from aggregation and disaggregation of the observations.

12. The increasing prevalence of both husbands and wives working raises the issue of the direction of causality when considered in conjunction with soaring housing costs particularly for single-family dwellings. Certainly, two or more workers within a household may be a prerequisite for homeownership as we enter the 1980s. On the other hand, escalating cost-price realities have, to some degree, been a result of a "demand-pull" phenomenon as asserted by Downs (1978). That is, the income resources commanded by two-worker families have served to create a strong demand function, pulling the price level upwards.

13. Annual Housing Survey data show rents increasing faster than both renter and owner household income. In contrast, Consumer Price Index data show rent increases lagging behind income. The difference centers on the fact that the CPI rental component gauges price changes attendant on housing of constant quality, while the Annual Housing Survey reflects the rents of the actual pools of housing. For a discussion on the point, see Sternlieb and Hughes (forthcoming).

REFERENCES

DOWNS, A. (1978) "Public policy and the rising cost of housing." Real Estate Review 8: 27-38.

JOHNSON, B. L. (1979) "Changes in marital and family characteristics of workers, 1970-78." Monthly Labor Review (April): 49-52.

RAWLINGS, S. (1978) "Perspectives on American husbands and wives." Current Population Reports, Special Studies Series 77: 23 Washington, DC: U.S. Government Printing Office.

STERNLIEB, G. and J. W. HUGHES (1979) "New dimensions of the urban crisis," in "Is the Urban Crisis Over," Hearing Before the Subcommittee on Fiscal and Intergovernmental Policy of the Joint Economic Committee, U.S. Congress. Washington, DC: U.S. Government Printing Office.

--- (forthcoming) Housing, Problems and Prospects. New Brunswick, NJ: Center for Urban Policy Research, Rutgers University.

THOMPSON, W. R. (1978) "Aging industries and cities: time and tides in the Northeast," pp. 144-152 in G. Sternlieb and J. W. Hughes (eds.) Revitalizing the Northeast: Prelude to an Agenda. New Brunswick, NJ: Center for Urban Policy Research, Rutgers University.

U.S. Bureau of the Census (1979) "Money income in 1977 of families and persons in the United States." Current Population Reports, Series P-60, No. 118. Washington, DC: U.S. Government Printing Office.

--- (1971) "Income in 1970 of families and persons in the United States." Current Population Reports, Series P-60, No. 80. Washington, DC: U.S. Government Printing Office.

U.S. Department of Commerce, U.S. Bureau of the Census (1978) General Housing Characteristics for the United States and Regions: 1976. Annual Housing Survey, 1976, Part A. Current Housing Reports, Series H-150-76. Washington, DC: U.S. Government Printing Office.

6

Residential Mobility and Urban Policy: Some Sociological Considerations

WILLIAM MICHELSON

☐ A SOCIOLOGICAL ANALYSIS of any phenomenon involves consideration of the appropriate level of focus. The implications of residential mobility demand qualification: For whom? For what? In this chapter I shall indicate the inadvisability of lumping the many varying types of residential mobility together, expecting any common sets of dynamics or consequences. Effects on individual movers are not, in aggregate, as stressful as observers were once likely to believe, and I shall cite several reasons for this. I shall proceed to discuss two aspects of mobility that are indeed problematic and which have policy implications; while dependent on the values and experience of individuals, they are solved only at higher levels of social and political organization.

MOBILITY AND IMPACT

"Breathing may be dangerous for your health," says an antismoking poster. This admonition is, of course, intended for a specific situation. Breathing may indeed be dangerous in a smoke-filled room, or in an asbestos mine. And all of us breathe. Most of use move, too. The circumstances of these moves are highly varied—from international moves to intramural changes of apartment.

Yet, movement itself has been seen, at best, as an unsettling experience and, at worst, as undermining established order. Those changing residence have been viewed over the years as fomenting civil strife, escalating needs for urban welfare services, becoming individually alienated, and harboring great stresses in the process. However, "in an average year, 36 or 37 million Americans move from one home or apartment to another" (Long, 1977: 914); it is not evident that they—or others before and after them—are significantly affected by the process. Research studies do not uncover major, consistent social implications from moving, even among the most dramatic types of mobility (Sanna, 1969).

Many previous stereotypes about mobility have been exploded by careful analyses. For example, Tilly laid to rest the notion that urban migrants were responsible for the European civil disorders of the nineteenth century, noting that there was "a positive connection between organization and conflict . . . rather than individual disorientation or malaise from exposure to the modern city" (Tilly, 1974: 103). Long (1977) showed that the great welfare caseloads of the mid-twentieth century reflected applications largely from nonmigrants. Research indicates that the connection between success in school and mobility is highly dubious, even for those moving in midterm, when proper controls are inserted (Barrett and Noble, 1972). Studies throw similarly cold water on alienation hypotheses (Butler et al., 1973; Jones, 1973) and expectations of stress (Butler et al., 1973; Tennant and Andrews, 1976).

When focusing on individuals and their moves, why do we see such weak aggregate effects? There are several reasons. First, most moves are relatively short. According to Butler et al. (1969), four-fifths of all American moves are within metropolitan regions, half with the central city as destination. In Canada, 47 percent of the population in 1971 (over five years of age) had moved in the five years since 1966. However, of these movers, slightly more than half (51 percent) stayed within the same city, and 82 percent remained within the same province (Statistics Canada, 1977: 8). Simmons (1974) focused on the more microscopic scene in metropolitan Toronto. Among the 51 percent who moved, about 20 percent changed neighborhoods (many suburbanizing), but usually remained in the same sectors of the metropolitan area, not very far from their previous homes (Simmons, 1974). Butler's national sample in the United States showed that of the 33 percent of families that moved between 1966 and 1969, 25 percent stayed in the same neighborhood and 40 percent stayed in the same municipality (Butler et al., 1969).

Second, studies generally indicate that this large cohort of short-distance movers is usually acquainted with the locations to which they move due to previous travels and the associated perceptions of these locations. People become aware of areas within cities through either regular trips (for work or shopping) or occasional trips (to visit friends or relatives). In this way areas become familiar, and moves to them are not abrupt or bewildering (Gad et al., 1973; Lee, 1970). Even when regular trips do not lead to new locations, use of the automobile is a North American technique for becoming aware of areas and opportunities. People who have some choice seldom choose a "pig in a poke" (Clark, 1966; Michelson, 1977).

Third, movement is often cushioned by existing personal or institutional contacts. Stouffer's long-standing theory of intervening opportunities (1940) suggests quite rightly how many long-distance moves are a function of unparalleled economic opportunity, but leaves the impression that people follow work opportunities without the benefit of peer support. The "welcome wagon," by its existence, gives the same impression. Yet, studies show that poorer inmigrants or international migrants are often moving under the auspices of family or friends who preceded them: The new hosts frequently share (crowded) dwellings and provide marginal job information, but the cushioning effect is there (Tilly and Brown, 1967). Even guest worker and refugee migrations are increasingly cushioned, under government or government-supervised auspices. Corporate migration is facilitated by employers and colleagues; indeed, this is now so much the case that corporation use of nationwide realty transfer firms is threatening the existence of independent real estate agencies in some areas. Earlier work on new suburban areas, furthermore, suggests that social patterns there support the intercity mover, who usually transfers from one suburb to another (Whyte, 1956).

Fourth, although no one would deny dysfunctions accompanying moves, there are usually strong positive motives precipitating many sorts of moves. Reasons for moving are broken down into pushes and pulls, the former usually negative and the latter positive. We often focus on the former without paying explicit enough attention to the latter. Many American families engaged in so-called "white flight" are often cited as examples of reactions to real or imagined difficulties with interracial contact, but their numbers would not be so large nor their motivations so strong were the moves not to homes and neighborhoods to which they have been aspiring all along (Frey, 1979).

In sum, reasons for moves are many and varied. Some are purely economic (change of job or speculation in housing) while others are without economic motivation (divorce or health). Some are voluntary (upgrading housing); others are involuntary (renewal or eviction). Some are based on purely demographic conditions (moves to larger apartments), while others reflect lifestyle considerations (moves to particular suburbs). Whatever the reason, there are usually positive factors underlying moves.

Finally, major life events—even if financially taxing, as is the case regarding housing—are not necessarily distressing. Australian researchers took the well-known Holmes and Rahe (1967) scale of potential life stressors, on which residential mobility ranked high, and empirically differentiated the extent they were distressing, as compared with simply representing a life change. The greater the move, the more stressful it is; international moves were ranked as moderately stressful. However, no form of move was seen as highly distressing, and local moves were among the least distressing of 67 types of life change (Tennant and Andrews, 1976).

The above arguments are not meant to suggest that all people moving are without stress, or that some types of moves are not more stressful than others, or that there are not potential intrafamilial differences in consequences from moving. The point is that there is no good evidence to support expectations that moving per se is a highly disturbing process. This must not lead us, however, to ignore those aspects of moving which are significant at the individual level and which raise questions of urban policy. I shall deal with two of these at length.

FORCED MOBILITY

Forced, or involuntary, mobility takes more than one form, and it applies on both the international and local scale. At the latter level, it applies to the tenant required to vacate, contrary to intention or expectation. It applies to the residents of buildings expropriated for urban renewal, tenant and owner alike. In any case, the push is the significant factor in the move, and the usually positive pull factors lack some combination of knowledge, time, and motivation for adequate germination.

The category of forced movers can be expanded, if one extends the logic to intrafamilial differences in motivation. In some families there are differential interests relating to potential moves. A corporate or military transfer, for example, traditionally originated in the concerns of the male breadwinner. Under circumstances when the wife counted herself as a loyal part of the corporate team and equivalent housing and social contacts were provided at the other end, serious or lasting disruptions were not prevalent—though wives have been noted as more prone to such discontinuities as occur (for example, McAllister et al., 1972; Gutman, 1963). Recently, however, many more women (including mothers of young children) have paid employment, often in career positions. Separation and divorce rates have also grown, in large numbers of cases involving children. Both trends make major moves more difficult for the partner (or ex-partner) not directly involved in the reason for the move. These trends increasingly serve to deter the making of some moves. Executives, for example, appear more resistant to facile and frequent transfers; but, when undertaken, these moves bring about more involuntary upheaval for the second person.

Children have always been a source of concern in residential mobility, though not usually a deterrent. Many short-distance moves, particularly to suburbs, are made "in the interest of children," although we have frequently failed to understand the actual needs children have for activity opportunities in newly built areas (for example, Michelson and Roberts, 1979; Larkin, 1979). Nonetheless, children's interests have not been a paramount concern in many types of moves. Although the children's rights movement has been less rigorous and of shorter duration than has the women's movement, there is currently the potential for still another part of the family to become aware that their investment in a given milieu can be arbitrarily disrupted by the volition of another member of the family.

Evidence from research suggests that, all else being equal, involuntary movers are far more likely to be unhappy with their new surroundings (Egerö, 1967) and subject to potentially serious psychosomatic disorders (Fried, 1973). Indeed, Fried's in-depth study of residents of the former West End of Boston, indicating various symptoms of "grief" (including a form of colitis) over several years after a forced move due to urban renewal, contributed to revised perspectives on the nature of renewal in the United States.

Several policy aspects of forced mobility deserve attention, not the least because large numbers of persons may be involved. Rossi, for example, estimated that as much as 40 percent of moves may be considered involuntary (Rossi, 1955). First, should steps be taken to minimize forced mobility? Nations like Britain and Sweden, for example, have introduced certain forms of tenure for renters to protect them against the personal effects of arbitrary actions by landlords. Such regulations have had negative side effects with respect to the size of the rental market; but it should not be beyond the ingenuity of policy makers to develop schemes which also protect legitimate landlord interests. North Americans probably agree on the inhumanity of mass expulsion of ethnic populations in other nations, but have our own private and public sector institutes adequately considered the implications of their own regular transfer policies?

Second, given the fact of forced mobility under certain circumstances, have policy makers at various levels come to terms with optimal ways to cushion the practical and emotional problems of forced relocatees? Neighborhood continuity, for example, is important for the functioning of senior citizens (Michelson, 1970). When, through various circumstances, they are forced to move, are older persons given every form of assistance to remain in familiar areas? Do institutions help to provide continuity to the careers and plans of other family members among households they transfer? Do public and private refugee sponsors deal adequately with the need for support well beyond the arrangement of jobs and household necessities for those undergoing the shock of instant international transplantation?

Policies developed for the good of the whole should not, in a democratic society, be arbitrarily applied to minorities. The starting point in understanding effects is, in any case, an appreciation of how the individuals in a household encounter and experience the change. However, amelioration lies at higher levels, both regarding prevention and, if necessary, post hoc support.

BLOCKED MOBILITY

I have already argued that people employ a diversity of criteria for choosing and then evaluating housing. Such criteria vary according to the circumstances of the move. Satisfaction or dissatisfaction in housing does not occur in a vacuum, but with respect to specific standards which people establish and maintain. Understanding the appropriateness or success of moves for people, then, requires an understanding of the criteria these people have utilized—and why.

The range of evaluative criteria from which to choose means that there is always a potential discrepancy between people's reasons for doing things and those outside observers assume they have. Economists, for example, suggest that if people buy or rent something, they must want it; but this is an inaccurate assumption with respect to housing in an imperfect market. Architects assume that if housing were to contain the proper assortment of "user-needs" (that is, functional attributes of the dwelling unit), occupants would be satisfied; but this view is also overly selective and static in its treatment of what people want and for how long. To complicate matters, the criteria people use are changeable from one time to another. A longitudinal housing study we conducted in Toronto[1] resulted in the delineation of what we called a "family mobility cycle" and some empirical evidence of its implications regarding mobility and satisfaction in housing (Michelson, 1977).

The family mobility cycle suggests that people's pattern of housing choice over time and the criteria underlying them are not totally random, inchoate, or subject simply to the eventual development of stress in their existing residence, as the more sophisticated geographic theories of mobility suggest (for example, Brown and Moore, 1970; Clark and Cadwallader, 1973). While surely not uniform or lock-step, people's mobility behavior is oriented toward the achievement of culturally prescribed ideals in housing (Morris and Winter, 1978); and their view of moves and housing is predicated on where they stand in an overall process of intended mobility, together with the possibility of approximating what they consider their ideal.

The family mobility cycle suggests three highly general but heuristic stages in residential mobility. First, there is the *baseline stage.* This is either a family's first home together or their first in a metropolitan area which is new for them. This tautology takes on meaning in the current context because the family cannot judge from past experience what is optimal for their current circumstances. Hence, baseline residences are chosen on relatively simple grounds (such as location of work place). The second stage, *incremental change,* assumes that the residence chosen in the baseline stage is not the ideal to which families aspire and that there is a period of variable (sometimes indefinite) length before people find it practical to achieve or approximate their ideal in housing. During this stage, people adjust to desires to fine-tune their housing beyond what they found upon entrance to the market while still not getting their ultimate wishes. Such fine-tuning, even if it does not terminate further family mobility, does allow the focus to turn to other sources of gratification or pragmatic efficiency (for example, recreational facilities or access to urban amenities) which bear no necessary resemblance to the criteria most people place on their ultimate choices of residence (such as security of tenure, self-sufficiency, land, neighborhood quality, or freedom of alteration). However, the ideal may never be achieved for some families, although its perceived possibility is important in giving hope, direction, and a feeling of relative flexibility to families in earlier stages. Because there are so many criteria applicable to housing, no single unit is likely to satisfy all criteria. Therefore, when families *approximate the ideal* (the third stage), they are emphasizing their most strongly held criteria and very likely compromising other criteria they previously employed. For example, the family moving from a luxury downtown apartment to their suburban dream typically sacrifices access and on-site recreation facilities, but gains a package of attributes even more desirable.

Thus, viewing family mobility and its underlying criteria as a cycle not tied directly to demographic change or dissatisfaction helps to explain why the same family may utilize apparently different criteria for housing choice at different times. It also helps us to understand why mobility is so high, and considered so regular, whenever opportunity permits.

There may never be a societal consensus on what form of housing is most desirable, due to the many differences in families' objective circumstances and their sociocultural backgrounds. Yet, in most societies, there is a widely shared view of desirable homes and neighborhoods. In North America, the modal cultural ideal remains the single-family house (Dillman et al., 1979), although its optimal location varies between central-city and suburb. Under the family mobility perspective, the possibility in North America of eventually gaining a private home helps determine what people demand from a new place of residence and the basis of their opinion of it. This can be represented, among other ways, in the form of a housing genealogy. As Figure 6.1 indicates, it is straightforward to chart mobility relevant to housing type. Looking at residential mobility this way, however, not only clarifies the timing of moves but also provides clues as to the nature of family demands on housing at any point through the provision of a long-term context in which a given move or form of residence can be placed.

However, if mobility is so frequent and normal, relating to an expected pattern of varying goals to be served by it, what are the consequences of an inability to move? An analysis of the Toronto data (Michelson, 1977), indicated that families in high-rise apartments said that satisfaction or dissatisfaction with the housing unit rested not with the units' functional characteristics but with respondents' perceived ability to move to single-family housing in the future, providing that was their ideal. People living in apartments were unhappy if they desired to move to a house in the future but were apparently unable to do so, while others in the same buildings were very happy—if they felt able to move or if, in a small minority of cases, the high rise was their ideal. In the case of those with blocked mobility, their condition did not permit them that mental luxury of judging the apartments on immediate criteria which the units could satisfy (for example, efficiency); instead, they employed their most strongly held values (which the others could "store" in their minds for a later time), finding apartments wanting.

The paradigm suggested above serves to clarify why so many people who are satisfied with the housing they choose nonetheless move again soon. In our study, satisfaction with housing in no way predicted future mobility. The strength of aspirations, a powerful force, was shown in the fulfillment of originally stated intentions during the four and a half years of the study. Although such a perspective was fruitful in assessing market behavior and its consequences for middle-class persons,

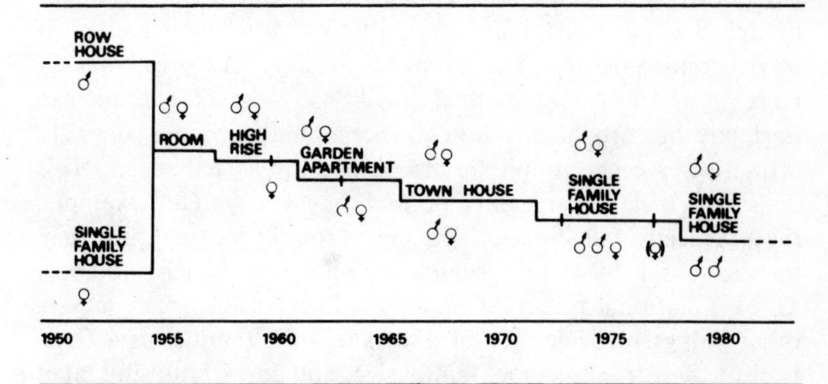

Figure 6.1 Hypothetical Housing Genealogy of a Family in Child-Bearing Years

its implications are even more telling for the poor, in housing in the private or public sector. For the poor, blocked mobility according to cultural norms is at its greatest. The family mobility cycle perspective and the accompanying evaluative dynamics coming from blocked mobility help to account for the intensity of dislike for projects which have been promulgated as functionally adequate and sufficiently spacious but which differ radically in appearance and atmosphere from cultural ideals (Kennedy, 1975).

Many argue logically that the continuity of the single-family house as the ideal reflects traditional economic realities of North American real estate markets. It is the best investment (geared to tax breaks), and people want it on those grounds. An extension of this argument is that people will increasingly adopt other forms of housing with equivalent tenure and economic aspects as parallel ideals. For this reason, subsequent analyses were conducted in Sweden, where the rules and alternatives form a very different context. Sweden is an excellent nation in which to explore the family mobility cycle further because many rules of the "housing game" differ from those in North America.

First, the majority of new housing units, particularly apartments, have been built by nonprofit organizations—cooperative housing associations and local municipalities. There is a high percentage of owned apartments on the cooperative apartment model. Second, apartments are built to a very high standard of service and amenity. The number, quality, and proximity of recreational and commercial facilities is greater than can normally be provided to homeowners. Landscaping is generally attractive. Access to public transportation is enhanced. This does not make apartments perfect in all ways (for example, they typically fall short in size and in soundproofing), but they possess a set of virtues typically lacking in North America. Third, a national policy of income redistribution which, among other things, provides rental allowances for families as a function of their total income, family size, and cost of housing takes away some of the worries of arbitrary rent increases among nonowners. Fourth, the large share of the market held by nonprofit landlords has been consciously used to keep down the monthly costs of rental and nonprofit housing in other segments of the market. Fifth, during the decade 1963-1972, about 80 percent of the one million new housing units built were in multiple-family dwellings, which became familiar to a large segment of the Swedish population. Finally, although there are still economic advantages to homeownership in Sweden, focusing on tax deductions and long-term appreciation of values, the overall picture is not as skewed toward homeownership as in some other nations.

Sweden therefore assumes interest as a setting in which to test whether families in different types of housing are in different stages of a family mobility cycle and whether, if present, such a cycle assumes the same dimensions as those seen in Canada. If these questions are answered positively, it suggests that some of the noneconomic dimensions of housing are indeed crucial in housing market behavior.

THE SWEDISH EXPERIENCE

An original survey of housing and mobility of families renting and owning similarly designed and located low-rise garden apartments in Malmö was undertaken as part of a study of user attitudes toward and uses of furniture.[2] Questions covered furnishing, demographic characteristics, residential history, and rationales for housing choices. Followup interviews were held within the following year (that is, one and a half to three years after the family had moved to its new housing), to assess any subsequent movement and attitudes toward repeat mobility and ideal housing. Complete information from the male head of household was received during the main interview from 53 men in the owned apartments (Almvik) and from 39 men in the rental apartments (Holma). Follow-up interviews were completed with 42 and 24, respectively.

It is possible with this information to explore the family mobility cycle. Parents of families in the two projects were all in their twenties, with similar distributions in social class, family size, and age of children. The apartments cost somewhat more to rent than to own on a monthly basis, although the owned apartments required a capital investment. Household incomes were higher among the owners–about 75,000 kronor per year compared with 60,000 among the renters. In consequence, the renters were more likely to receive a housing allowance than were the owners (though about a third of the latter also receive such an allowance; Lindström and Åhlund, 1977).

According to the family mobility cycle perspective, we would expect that (1) those moving to Almvik have made more moves in the recent past and (2) have different reasons for selecting Almvik than do those selecting Holma. Nonetheless, (3) the two groups should share the same set of goals in housing (owned, detached houses), and (4) the particular reasons for selecting Almvik and Holma fit into a coherent long-range pattern of housing choice.

Since the men were in their late twenties, a table was made of the number of previous moves they had made in the preceding 10 years (since 1966). Table 6.1 shows that residents of Almvik had made more moves, as expected. The mean number of previous moves among those in Almvik was 2.3, compared with 1.79 in Holma. However, only 33.5 percent in the former area had not made more than one previous move, as compared with 51.2 percent in the latter.[3]

TABLE 6.1 Previous Moves Since 1966 Among Male Heads of Household in Two Malmö Housing Areas (percentages)

Number of Moves	*Almvik*	*Holma*
0	7.5	17.9
1	26.4	33.3
2	32.0	17.9
3	9.4	17.9
4	17.0	7.7
5+	7.6	5.1
	99.9	99.8
N	53	39

Table 6.2 indicates the main reasons people moved to Almvik/Holma. As expected, these reasons differ between the two projects. For a third of the families, the rental accommodation was the first place of residence of a new household, compared with 15 percent among those choosing the owned apartments. In contrast, the latter, with more experience, selected their housing with greater attention to the finer design details of the housing; about a third of them mentioned design aspects of the housing, compared with 5.1 percent of the renters—this despite the great similarity in the buildings. Getting sufficient space was mentioned frequently by both groups—by 33.3 percent of those in Holma and by 26.4 percent in Almvik.

That such differing rationales reflect position in the family mobility cycle rather than objective differences in the housing is shown in Table 6.3. This table pools the men from both projects, in each of which are found people with varying degrees of previous mobility despite the differing central tendencies. The rationale for each of up to four successive moves since 1966 was recorded according to whether it was the first move during this period, the second (if relevant), the third, or the fourth. In the case of men who had moved often, these data do not include the Holma/Almvik move—only the first four since 1966. Almost by definition, leaving the original family home and setting up a new family structure are rationales for choosing

TABLE 6.2 Primary Reasons Why Male Heads of Household Moved to Two Malmö Housing Areas (percentages)

Reasons for Move	*Almvik*	*Holma*
Space	26.4	33.3
Leaving parental home	7.5	2.6
New ménage	15.1	33.3
Job change	1.9	10.6
Considerations for children	5.7	0.0
Design of housing	32.1	5.1
Dissatisfaction with previous home	11.3	12.8
No answer	0.0	2.6
	100.0	100.3
N	53	39

TABLE 6.3 Primary Reason for Move by Sequence of Move in Last 10 Years (percentages)

Reason for Move	*1st Move*	*2nd Move*	*3rd Move*	*4th Move*
Space	13.3	27.2	26.4	26.7
Leaving parental home	27.8	11.1	5.7	3.3
New ménage	25.6	24.7	13.2	16.7
Job change	15.6	8.6	9.4	13.3
Economic factors	1.1	0.0	3.8	0.0
Considerations for children	0.0	0.0	3.8	0.0
Design of housing	11.0	22.2	26.4	23.3
Dissatisfaction with previous home	3.3	3.7	11.3	13.3
No answer	2.2	2.5	0.0	3.3
	100.0	100.0	100.0	99.9
N	90	81	53	30
No migration history given	2	2	2	2
Not applicable	0	9	37	60

housing which occur early and then decline in importance. While this occurs, space and design are not paramount (although surely not irrelevant). Job location and changes thereto are a relatively constant factor among the mobility of families, not diminished by the parallel setting up of new households. With subsequent moves, however, space and design assume the greater importance shown in Almvik (Table 6.2). People also start to move in reaction to inadequacies in their existing housing; they examine the details of their prospective new housing more closely than in earlier situations where the main goal was simply a suitably located roof over one's head.

The sequence of mobility represented by Holma and Almvik is underscored by the subsequent mobility of these families. Among those who were interviewed in the follow-up, 42 percent of those in the rental apartments had moved again (30 percent inside Holma and 70 percent elsewhere), compared with only seven percent who had invested in Almvik. Unfortunately, there is no record of their choice of housing type.

Nonetheless, the goal-oriented aspect of the family mobility cycle is shown by the extremely similar mobility intentions of those who had not moved again. About 80 percent of both cohorts expressed the intention to move again within five years. According to about two-thirds of both groups, they hoped such moves would be to a single-family home. Only about one out of five in Holma and one in six in Almvik desired improved versions of their current housing type and tenure, with the remainder wishing one or another form of predominantly low-rise accommodation.

That the aspiration for detached houses does *not* reflect the economic aspects of home ownership primarily is indicated by answers to the following question: "It is claimed that there is an economic advantage to living in a single-family home compared to a multiple-family dwelling. If this advantage weren't found, would it be equally good to live in an apartment as in a single-family home?" About 70 percent of both groups answered "no" to this question.

It is interesting to note, however, that a (not overwhelming) majority of those in the owned apartments who answered "yes" to this question do not intend to move in the next five years, a sharp contrast to those in Almvik answering "no" and to all those in Holma. This gives some evidence of the size of the prime market for owned apartments—in contrast to the larger number of persons who reside there for a period of time but without seeing this as an ultimate home. Table 6.4 gives these

TABLE 6.4 Intention to Move in Next Five Years Among Male Heads of Household in Two Malmö Housing Areas (percentages)*

	Almvik		*Holma*	
Intention to Move	*Apt. as Good*	*Not as Good*	*Apt. as Good*	*Not as Good*
Yes	46.1	89.9	80.0	84.6
No	53.9	10.3	20.0	15.4
	100.0	100.0	100.0	100.0
N	13	29	5	13

*By agreement that, except for economic factors, apartments are as good as houses.

breakdowns. Such statements, expressed over and over and joined by only a few mentions of economic aspects like "cheaper" and "the rent," are virtually indistinguishable from those commonly given by North American respondents.

The Swedish data, therefore, add some degree of confirmation to a perspective on family mobility studies under different rules, suggesting the generality of the noneconomic components of the situation. The prevalence of goal-oriented mobility, moreover, underlines the seriousness of blocked mobility among others in society.

As with forced mobility, the factors behind blocked mobility are at a level above and beyond the individual or family. A light but not unduly flippant summary of the Toronto findings is that "the most effective way to increase satisfaction in high-rise apartments is to increase the accessibility of single-family houses." If the Toronto market had a different shape, the criteria upon which many residents judged their high-rise apartments would change, as would their satisfaction. However, changing the market so as to unblock mobility for many would involve taking actions to lower prices, probably including some emphasis on the production of basic (small and simple) houses, the logical attraction to blocked apartment residents. In the Canadian context, this would require politically sensitive steps to discourage or eliminate unproductive speculation and to speed up the red tape slowing down the construction of housing. These are macro solutions, based on problems understood in relation to the expectations and orientations of individuals.

Poorer people's mobility is blocked when housing assistance programs allocate particular housing units rather than offering grants based on housing costs, income, and family size. The latter approach, implemented in Scandinavia, assumes that those needing assistance do not, by token of their need, necessarily fit into a separate sector of humanity with separate norms and aspirations. In this regard, the literature from forced mobility joins to support this method of unblocking normal movements: If people are given the means for making a choice, housing for poorer people will automatically be "better" because its occupants will indeed have made a choice and can employ situationally appropriate means of evaluation, because they too could move again in the future.

CONCLUSION

If we succeed in curtailing forced mobility and in removing obstacles to blocked mobility, we remain with a society characterized by high rates of mobility. There are many market and institutional mechanisms designed to facilitate the act of residential change; however, generally lacking are universalistic practices making possible the kind of postmove personal supports which were available in stable communities. I do not advocate a return to "rule by gossip" or to geographically limited career mobility. Yet, is it necessary that young parents, having moved from relatives, suffer from an absence of informal advice about child development and problems, or that single parents lack companionship and assistance for the same reasons?

Planners once thought that the small-town social atmosphere could be recreated in cities through the physical design of new suburban neighborhoods (Perry, 1966). Although studies have indicated some physical determinism of most likely potential contacts under circumstances of class and life-cycle stage homogeneity (Michelson, 1970), there is no evidence of a return to unified local communities. Neighborhoods can be spatially proximate and unified, but no less often they are completely aspatial or of limited social significance (Wellman, 1979).

Can public policy respond in such a way as to provide supportive information and services otherwise lost? Can it foster cooperation and self-help among neighbors? Whatever the arrangement or form of supportive structure, one cannot assume it can maintain itself on the basis of personal friendships and solidarity because, one by one, families will disappear from the network due to mobility. One currently attempted vehicle is placing responsibility for local maintenance and recreation into the hands of community associations with mandatory membership. Research will be necessary to indicate whether this or other vehicles provide social support among neighbors as well as its more perfunctory purposes. What is reasonably certain is that the person arriving on the scene from elsewhere and the person left behind by the rest of his or her neighbors cannot arrange the functions and services of a collectivity as individuals.

Thus, although mobility per se is not evidently destructive to those involved, there are anomalies and residues which, though understandable from the point of view of the individual and family, require attention at higher levels in the interests of an optimal and attainable general quality of life.

NOTES

1. Work for the study was conducted under contract to Central Mortgage and Housing Corporation and with supplementary grants from The Canada Council. The auspices of University of Toronto's Centre for Urban and Community Studies were extremely helpful and pleasant during the conduct of this work, during the period 1969-1975.

2. The data for this study were gathered with the aid of a Canada Council Leave Fellowship, and a small grant from the University of Toronto facilitated the analysis. Malin Lindberg and Willem van Vliet assisted in data collection and analysis, respec-

tively. The Malmö data were gathered as part of a larger study on household furnishing organized and conducted by Birgitta Lindström and Owe Åhlund of the Institute for Building Functions Analysis, Lund Institute of Technology, under contract to Konsumentverket; I am grateful for their generous cooperation. Karin Boalt and Göran Lindberg were extremely helpful in making this research possible as part of a sabbatical year in Lund. These data were taken from a larger paper (Michelson 1980).

3. Similar results are obtained if lifetime moves are considered. Thus, the arbitrary ten-year period is not unfair.

REFERENCES

BARRETT, C. L. and H. NOBLE (1972) "Mothers' anxieties versus the effects of long distance move on children." Journal of Marriage and the Family 9: 181-188.

BROWN, L. A. and E. G. MOORE (1970) "The intra-urban migration process: a perspective." General Systems 15: 109-122.

BUTLER, E. et al. (1973) "The effects of voluntary and involuntary residential mobility on females and males." Journal of Marriage and the Family 10: 219-227.

--- (1969) Moving Behavior and Residential Choice: A National Survey. Washington, DC: National Academy of Sciences.

CLARK, S. D. (1966) The Suburban Society. Toronto: University of Toronto Press.

CLARK, W.A.V. and M. CADWALLADER (1973) "Locational stress and residential mobility." Environment and Behavior 5: 29-41.

DILLMAN, D. A. (1979) "Influence of housing norms and personal characteristics on stated housing preferences." Presented to the annual meeting of the American Sociological Association.

EGERÖ, B. (1967) Ny bostad i ytterstad. Goteborg: Sociologiska Institutionen, Goteborgs Universitet.

FREY, W. H. (1979) "Central city white flight: racial and nonracial causes." American Sociological Review 3: 425-448.

FRIED, M. (1973) The World of the Urban Working Class. Cambridge, MA: Harvard University Press.

GAD, G., R. PEDDIE, and J. PUNTER (1973) "Ethnic differences in the residential search process," pp. 168-180 in L. S. Bourne, R. MacKinnan, and J. Simmons (eds.) The Form of Cities in Central Canada: Selected Papers. Toronto: University of Toronto Press.

GUTMAN, R. (1963) "Population mobility in the American middle class," pp. 172-183 in L. J. Duhl (ed.) The Urban Condition. New York: Basic Books.

HOLMES, T. H. and R. H. RAHE (1967) "The social readjustment rating scale." Journal of Psychosomatic Research 12: 213ff.

JONES, S. B. (1973) "Geographic mobility as seen by the wife and mother." Journal of Marriage and the Family 10: 210-218.

KENNEDY, L. W. (1975) Residential Mobility as a Cyclical Process: The Evaluation of the Home Environment Both Before and After the Move. Doctoral dissertation. Toronto: University of Toronto.

LARKIN, R. (1979) Suburban Youth in Cultural Crisis. New York: Oxford University Press.

LEE, T. R. (1970) "Perceived distance as a function of direction in the city." Environment and Behavior 2: 40-51.

LINDSTRÖM, B. and O. ÅHLUND (1977) Möbler och Möberling i Unga Hushåll. Stockholm: Konsumentverket.

LONG, Larry H. (1977) "Migration and resettlement services," pp. 914-919 in Encyclopedia of Social Work. Washington, DC: National Association of Social Workers.

McALLISTER, R. J. et al. (1972) "Residential mobility of blacks and whites." American Journal of Sociology 77: 445-456.

MICHELSON, W. (1980) "Residential mobility as a dynamic process: a cross-cultural perspective," in C. Ungerson and V. Karn (eds.) The Consumer Experience of Housing. London: Gower.

--- (1977) Environmental Choice, Human Behavior, and Residential Satisfaction, New York: Oxford University Press.

--- (1970) Man and his Urban Environment: A Sociological Approach. Reading, MA: Addison-Wesley.

--- and E. ROBERTS (1979) "Children and the urban physical environment," in W. Michelson, S. V. Levine, and A. R. Spina (eds.) The Child in the City: Changes and Challenges. Toronto: University of Toronto Press.

MORRIS, E. and M. WINTER (1978) Housing, Family and Society. New York: John Wiley.

PERRY, C. (1966) "The neighborhood unit formula," pp. 94-109 in W.L.C. Wheaton, G. Milgram, and M. E. Meyerson (eds.) Urban Housing. New York: Free Press.

ROSSI, P. (1955) Why Families Move. New York: Free Press.

SANNA, V. D. (1969) "Immigration, migration, and mental illness," pp. 291-352 in E. B. Brody (ed.) Behavior in New Environments. Beverly Hills, CA: Sage.

SIMMONS, J. W. (1974) Patterns of Residential Movement in Metropolitan Toronto. Toronto: University of Toronto Press.

Statistics Canada (1977) Perspective Canada II. Ottawa: Ministry of Supplies and Services.

STOUFFER, S. A. (1940) "Intervening opportunities: a theory relating mobility and distance." American Sociological Review 5: 845-967.

TENNANT, C. and G. ANDREWS (1976) "A scale to measure the stress of life events." Australian and New Zealand Journal of Psychiatry 10: 27-32.

TILLY, C. (1974) "The chaos of the living city," pp. 86-108 in An Urban World. Boston: Little Brown.

--- and H. C. BROWN (1967) "On uprooting, kinship and the auspices of migration." International Journal of Comparative Sociology 8: 139-164.

WELLMAN, B. (1979) "The community question: the intimate networks of East Yonkers." American Journal of Sociology 5: 1201-1231.

WHYTE, W. H., Jr. (1956) The Organization Man. Garden City, NY: Doubleday.

7

Housing Market Search: Information Constraints and Efficiency

TERENCE R. SMITH and W.A.V. CLARK

□ SOCIAL SCIENTISTS have been increasingly interested in the way in which individuals and households make decisions. The questions of how choices or decisions are made and the relative role of individual and institutional variables influencing those choices have been central components of much social science research. Even a cursory review of the literature in economics and psychology yields a large number of articles which focus on human judgment and decision processes both in theoretical and applied settings. In any review of decision-making, information emerges as a fundamental component of individual choice, and there are a number of detailed reviews which analyze the role of information and information use. It is the intent of this chapter to analyze the role of information in the specific context of housing choice and to extend this discussion to include the policy implications of information use.

The housing market provides an excellent example of a decision-making environment in which there are several distinct channels of information and in which considerable amounts of resources are expended on the acquisition of spatial information. The process of housing market search can be viewed as a sequence of decisions concerning the allocation of resources for the purposes of obtaining information. The information which is obtained constitutes the input to further decision-making

AUTHORS' NOTE: *The research for this chapter was partially supported by National Science Foundation Grant SOC 77-27362.*

regarding both additional search and the allocation of resources for housing needs.

At the same time that the central role of information is recognized, the information environment is often poorly specified. Particularly in the context of the housing market we are dealing with an information environment which is large, complex, and uncertain. Although there are some data on the basic sources of information utilized by decision makers in the housing market, there is also confusion over exactly what data items are collected, how they are used, and their impact on those decisions. In addition, decisions concerning the use of the available information sources are dependent on the individual's cognitive model of the decision environment; but a feature which greatly complicates the study of search is that this cognitive model is subject to modification during the course of search.

An individual's search process is subject to constraints, and it is these constraints which have direct policy implications. A first set of constraints involves limitations on the resources available for obtaining information. These resources may be needed to overcome costs of transport, child care, and even of absence from work. A second set of constraints involves limitations on an individual's ability to model the decision-making environment and to process information. A third set of constraints, of particular interest for the present analysis, concerns restrictions on the flow of certain types of information. In general, the effects of such constraints on the structure and outcome of housing market search are not understood in much detail. Limitations on resources evidently limit the temporal and spatial extent of search (see Cronin, 1979a, 1979b; McCarthy, 1979), although it is not clear as to whether they directly lead to inefficient search in the sense of leading to less desirable outcomes. Constraints involving an individual's cognitive modeling and information processing are clearly important in relation to the efficiency of search. For example, a homebuyer's perceptions concerning the nature of available information sources and their interrelationships may effect both the efficiency and outcome of the search. Inadequate perceptions or misconceptions may lead to poor strategies of source usage and hence to the choice of a suboptimal vacancy. This in turn may result in more frequent relocation than is optimal in a given market, thereby wasting resources on additional search and transactions costs. There are often strong constraints on information flow in housing markets. One notable example involves restrictions on the public dissemination of MLS data by realty agents. These restrictions are sometimes supported by legal arguments. Another example relates to disclosure of loan disposition information by financial institutions.

Although the effects of information constraints on mobility are not well understood, it is evident that they provide a potential focus for housing policy. For example, policies might support subsidies for the provision of housing information for low-income households; they might encourage disclosure of MLS data; and, more generally, they might address the problem of agent fees and services. Before such policies can be investigated, however, it is essential that the constraints they seek to weaken be clearly understood, both in their mode of operation and in their effects. In this study we incorporate such information source constraints as information costs. Strong constraints have correspondingly high costs. The implication of such information source usage costs is that individuals will have imperfect information. As a result, their strategies of information source usage and the outcome of search will be influenced by such costs. For example, individuals who possess inadequate representations of the market may involve realty agents in fruitless search. Many agents claim that one reason for the widespread use of the six percent agent fee, despite inflated house prices, results from "wasted" search efforts. These high agent fees are a significant part of the transaction costs.

In addition to the problem that the information environment for housing market search is not well specified, it is also difficult to gather data on the nature of information use in housing market search. To obtain strong empirical results, large and detailed samples of individuals engaged in the decision-making process are required. Not only is this difficult at the present time, but until we have some better specification of the concepts involved in the decision-making process, it is likely that empirical observation would not necessarily be the best approach to the problem. If we accept the view that surveys are unlikely to yield useful insights on the use of information or, at best, elementary information on the sources of information, one viable alternative is to use a simulation approach. Assuming we need large numbers of cases under a variety of cost constraints, we can generate these results by allowing a number of "individuals" to engage in housing market search under the format of a model which describes the actions they will take, the beliefs they have, and their goals.

GENERAL INFORMATION AND SEARCH EFFICIENCY

The manner in which available information sources are used during housing market search depends on the individual's perceptions concerning the nature of the information system. These perceptions involve a knowledge of the existence of sources and the interrelationships among available sources. Given such knowledge, perceptions also involve the characteristics of various information sources, including the contents of messages from different sources and the costs of obtaining messages, as well as the reliability and timeliness of those messages.

For the purposes of understanding residential mobility, an important component of an individual's perceptions of the decision-making environment relates to the temporal and spatial distribution of vacancies by price and type. The information that modifies this cognitive map arises from a variety of sources, including newspaper listings, MLS data, realty agents, friends and relatives, as well as driving and walking around and examining vacancies. In general, only newspaper listings and MLS data provide an instantaneous view of the whole market. We term such information "general information." The degree to which general information is obtained and integrated into the individual's cognitive model will depend on such characteristics as the cost and reliability of messages from these sources.

Several questions of interest concern the use and effects of general information during the search process. We phrase our interpretation of information flow constraints as questions about the costs of obtaining messages. Other equally interesting questions relate to such characteristics as message reliability. A first question concerns the effect of the cost of general information on the spatial and temporal efficiency of the search. (Efficiency will be defined in detail later, but here it basically refers to the length of the search process.) A second question relates to the effect of the cost of general information sources on the use of other information sources. A third question involves the effect of the cost of general information messages on the outcome of the search. Answers to such questions will shed light on the role and effects of information constraints on housing search.

There are various ways in which these questions may be investigated, and all are subject to serious problems. Empirical observations on the information-seeking and information-processing behavior of individuals in housing market situations are fraught with difficulties of completeness and interpretation. Assessments of the effects of information constraints on search currently appear next to impossible to obtain. Attempts to model the search process are also beset with difficulties. Theoretical approaches, however, coupled with simulation or experimental work, may provide some useful insights into the problems.

In this chapter we examine the effects of the perceived costs of source usage on the spatial and temporal efficiency of search and on the sequential structure of the search process. In particular, we focus interest on the costs of general information. The investigation employs a simple model of decision-making in a housing market context as a basis for computer simulation of a homebuyer's actions. The results of the simulations are analyzed to determine the effects of perceived search costs on the efficiency of search. We then use transition graphs to examine further the effects of cost variation on the sequential structure of search, which provides a linkage between information cost and search efficiency.

THE RESEARCH CONTEXT

Most theoretical analyses of the mobility process and housing search have lumped the costs of information source usage into the general cost term of some disequilibrium model. Such models are based on the premise that the likelihood of relocation increases as the difference between the perceived benefits and the perceived costs of a move itself increases. The basic disequilibrium model is outlined by Hanushek and Quigley (1978) and by Smith et al. (1979). Refined measures of the benefits and costs of movement are given by Weinberg et- al. (1979) and Cronin (1979a, 1979b). While Hanushek and Quigley (1978) use the difference between the actual and equilibrium levels of housing consumption expressed as a fraction of current equilibrium demand, Weinberg et al. (1979) estimate the potential change in consumer surplus due to moving from an assumed demand function for housing services. Cronin (1979a, 1979b), on the other hand, makes specific assumptions about the form of the household utility function and computes the income equivalent variation (the amount of additional income necessary to make the household as well off with its current consumption of housing as it would be with its equilibrium consumption) of a move from actual to equilibrium consumption of housing.

For the most part, these studies do not pay specific attention to the different costs that enter into the disequilibrium model nor to their particular influence on the course and outcome of search. Cronin (1979b) and McCarthy (1978) considered effects in the search process that are presumably related to information costs: Cronin developed a logit model of mobility as a function of benefits, costs, psychological variables, and previous mobility in order to examine the usefulness of an economic measure of the benefits of moving for explaining the rate of search and the effect of race on search. In analyzing the data, Cronin noted that minority and nonminority households follow significantly different strategies during search, both with respect to sources of information and to the mode of transportation used during search. Significantly more nonminority households use newspapers, and significantly fewer minority households use their own automobile for search. He examined a number of indicators of search efforts, such as number of days searched, number of neighborhoods searched, and number of dwelling units searched.

McCarthy (1979) developed a three-stage model of search related to the general disequilibrium model; in particular, he discussed the determinants of various strategies of search and information source usage. He suggested two alternative strategies which the searcher may undertake: (1) minimize search costs or (2) maximize the likelihood of locating the best alternative available. The results of this study suggest that renters in particular often undertake a minimum cost strategy but frequently pay a premium for the unit ultimately selected. On the other hand, renters adopting a low-intensity search strategy, which relies on personal contacts, often obtain rent discounts, which we may associate with successful search.

While theoretical research has focused on the disequilibrium model and the degree to which general costs inhibit the probability of relocation, empirical survey research that specifically concerns the use of the housing market information system and its effects on the outcome of search has generally provided descriptions of the degree to which homebuyers utilize the various information sources available. The information sources generally considered include newspapers, walking or driving around, friends and relatives, and realty agents (see, for example, Rossi, 1955; Hempel, 1970; Barrett, 1973; Bettman et al., 1978). Measures of information source use include the usage rate (percentage of searchers using each source) and the location rate (percentage of searchers locating an acceptable vacancy with the source). A recent survey of these empirical analyses indicates a high degree of variability among the results of the studies and raises questions about their contribution to understanding the use of information (Clark and Smith, 1979).

The efficiency of the search process is largely unexplored either theoretically or empirically, and the sequential structure of information source usage has received inadequate attention. Only Hempel (1970; Hempel and McEwen, 1975) appears to have addressed the latter issue in any depth. He found a frequent sequence of source usage that involved newspapers, agents, and friends and relatives. The manner in which perceived costs or other source characteristics affect the sequential structure does not appear to have been investigated in detail, although McCarthy (1979) examined the relative cost of different sources on usage rate, finding that source costs are significant in determining search strategies.

Studies of information source usage in a housing market context have rarely involved formal modeling or controlled experimentation, which contrasts with studies in other areas of consumer decision-making. For example, Berning and Jacoby (1974) examined both the relative importance and the order of use of several information sources in new product evaluation, finding significant interaction between innovators and the sources used. In the psychological literature, there are reports of experiments in which subjects were allowed to purchase information from several sources in a sequential decision process (Kanarick et al., 1969; Levine and Samet, 1973; Levine et al., 1975). These studies indicate that a lower source reliability is associated with the purchase of more information and with a lower accuracy of the resulting decision. A limitation of these latter studies in relation to housing market information systems is that each channel was used to generate the same type of message, with only the cost and reliability of messages varying between channels.

Recently, Clark and Smith (1979) developed a model of a decision maker's use of information channels in housing search. The effects of cost variation on patterns of channel use and spatial search were investigated with simulation techniques. The results showed that cost variations led to four distinct patterns of channel use, while spatial search was dominated by the cost of obtaining general market information. The remainder of this chapter concerns extensions of this work and the implications for policy.

A MODEL OF INFORMATION USE

The effects of variations in perceived search costs are examined in terms of a decision-making model of individual behavior.

Although the model has been presented elsewhere in a different context (Clark and Smith, 1979), its basic structure is briefly outlined for the sake of completeness. The model may be described in terms of four major aspects of a decision-making process:

(1) the decision maker's beliefs,
(2) the decision maker's actions,
(3) the decision maker's goals, and
(4) the decision maker's mental processing.

It is to be emphasized that the model is elementary and of necessity involves assumptions that may be questioned. Our goal in developing and analyzing this model, however, is to obtain preliminary insights that will guide future research involving experimental work and theoretical analyses.

BELIEFS

The decision maker's subjective beliefs concern (a) the spatial structure of the market and the chances of finding an acceptable vacancy, and (b) the information channels and their characteristics. Both aspects of beliefs involve several components. In relation to market structure, the decision maker believes the housing market to be divided into a set of distinct subareas, each of which is relatively homogeneous with respect to the chance of finding an acceptable vacancy. We model the individual's beliefs concerning the chance of locating such a vacancy in terms of a probability distribution that depends on the subarea searched and the ability of the realty agent with whom the individual is searching. We assume that subareas are viewed as either good or bad, with good areas having a higher chance of giving rise to a vacancy. Similarly, agents are perceived as good or bad; a good agent has a better chance than a bad agent of locating an acceptable vacancy in a given subarea. The decision maker is uncertain as to which areas are good or bad, and we model this uncertainty in terms of subjective probability beliefs concerning whether each of the areas is good or bad.

$$\text{Probability (subarea i is good)} = \text{PGS (i)}, \; i = 1, n \qquad [1]$$

Such beliefs are revised during the course of search.

There are four channels of information, and the decision maker has a set of beliefs concerning each channel. The channels are:

C1: messages concerning which subareas are good.

C2: messages giving recommendations about realty agents.

C3: messages from the decision maker to an agent concerning the attributes of an acceptable vacancy.

C4: messages from the decision maker to an agent giving directives to search a specific subarea.

All channels are characterized by a cost per message (c_1, c_2, c_3, c_4), while C1 and C2 messages are also characterized by a message reliability. Each C1 message M is represented as a vector of submessages $M = (m_1, \ldots . m_n)$, in which submessage m_i indicates whether subarea i is good or bad. Submessages are assumed to be independent and characterized by a reliability R:

$$\text{Probability (submessage } M_i \text{ is correct)} = R. \qquad [2]$$

Such messages may be conceived as coming from newspaper vacancy listings.

Individuals may contact an agent with or without a recommendation, although obtaining a recommendation increases the chance of finding a good agent. Recommendations may be viewed as coming from friends and relatives. We model an individual's beliefs concerning the chance of finding a good agent in terms of two probabilities:

Probability (obtain a good agent by random selection) = PGAR [3]

Probability (obtain a good agent using a recommendation) = PGAM. [4]

On the other hand, whether an agent is good or bad is only revealed after a period of search with the agent. The third channel may be viewed as involving the initial process of dealing with an agent, while the fourth channel directs search to specified subareas.

ACTIONS

Other than not engaging in search activity (a_0) the individual decision maker may undertake a finite set of actions with respect to housing market search. These include:

a_1: Obtain a C1 message concerning the market subareas (collect general information).
a_2: Obtain a C2 message, or recommendation, concerning an agent (obtain a message about an agent).
a_3: Contact an agent about whom there is no prior information and search in a specified subarea (search randomly with an agent).
a_4: Contact a known agent and search in a specified area (search with a known agent).
a_5: Contact an agent concerning whom there is a C2 message and search in a specified area (search with an agent for whom there is a recommendation).
a_6: Withdraw from the market on locating an acceptable vacancy.

In any one period, the decision maker chooses only one of the seven actions, which involve either collecting information or searching in a specific subarea. The final action is the cessation of search activity (a_6) when an acceptable vacancy is found. At any stage in the search process, not all actions are feasible. For example, action a_5 specifies that the decision maker has a recommendation concerning an agent; hence, a_5 can only occur after a_2.

GOALS

We adopt a simple model of the decision maker's goals in which the individual maximizes expected utility. If U is the (constant) utility of locating an acceptable vacancy, and $\hat{s}$ is a given strategy for using the information sources, the individual chooses that strategy s such that

$$P(\hat{s})U - c(\hat{s}) = \max_{s} [P(s)U - c(s)], \qquad [5]$$

where P(s) is the probability of locating an acceptable vacancy with a strategy s and c(s) is a measure of the expected cost of the strategy. In order to obtain a more precise statement of a decision maker's goals, it is necessary to consider the mental processing of the individual.

MENTAL PROCESSING

The decision maker's mental processing involves a set of rules for belief modification during search and a set of rules for constructing and choosing search strategies. It is assumed that the decision maker updates beliefs concerning which subareas are good in a Bayesian manner in response to both general (C1) messages and unsuccessful search. Other beliefs are not modified. While it is known that individuals are not truly Bayesian, the assumption reflects the spirit of most observed behavior.

Concerning the choice of strategy, it is assumed that an individual obeys the following rules. First, a decision maker considers feasible sequences of actions that only involve the next two periods of search, reflecting constraints of the individual's computational abilities. In order for a two-action sequence to be feasible, it must contain only one action involving search with an agent (except for the "inactive" sequence $a_0 a_0$). Furthermore, contemplated actions must be compatible with past search activity (for example, it is not possible to search with an agent who is recommended, a_5, if no previous recommendation has been obtained, a_2). At the beginning of search, we define as feasible the following action sequences:

$$\begin{array}{l} a_0 a_0 \\ a_1 a_3 \\ a_2 a_3 \\ a_2 a_4 \\ a_3 a_0 \end{array} \qquad [6]$$

Other sequences are possible but are not considered at this time.

A second rule is that an individual chooses the two-action sequence $s_{ij} = a_i a_j$ giving rise to the highest expected utility. The expected utility is computed according to 4, with $P(s_{ij})$ being the probability of a successful search and $c(s_{ij}) = c_i + c_j$ the marginal cost of search. When considering the expected utility of the sequence of actions that involves obtaining C1 information as an intermediate step, the decision maker uses a Bayesian preposterior analysis in relation to the possible C1 message obtainable (see Morris, 1968). Again, we adopt a normative approach as an approximation to behavior.

A final rule is that on choosing an action pair $a_i a_j$, the individual only carries out the first action a_i of the pair. In response to the outcome of this action, a new set of feasible action pairs are contemplated, and the decision process is repeated.

To place the model in perspective, it is useful to consider the way in which it operates. Search begins with an individual having a set of prior beliefs but without an agent or recommendations concerning an agent. The individual then compares the expected utilities of the five initially feasible action pairs (5). From this set of action pairs, the individual chooses that combination giving rise to the highest expected utility. The first action of that pair is then undertaken. Next, beliefs are updated if either general information is obtained or if an unsuccessful search occurs in a specific area. In response to the outcome of this step, the individual reevaluates the expected utility of the set of feasible action pairs and again chooses that pair with the highest expected utility. If a recommendation is obtained in an earlier step, the individual may contemplate searching with the recommended agent. During this process, the set of feasible action pairs changes depending on whether the individual has an "unused" recommendation or whether the individual has had prior contact with an agent. The individual continues the selection of action pairs until either an acceptable vacancy is located or the individual withdraws from the market.

THE EFFECTS OF INFORMATION COSTS ON SEARCH

The results of a set of simulations based on the model are analyzed in terms of the efficiency and sequential structure of search. Our particular focus is on the role of the cost of general information, as it affects (a) the use of other sources of information, (b) the sequential structure of search, and (c) the efficiency of search. The first part of the analysis concerns the manner in which two measures of search efficiency vary with changes in the four perceived costs of search. Variations in the search costs lead to variations in the efficiency measures because they affect the sequential structure of search. The second part of the analysis concerns methods for representing the sequential structure and the effect of cost variations on this structure.

For the first part of the analysis, it is important to define "objective" measures of search efficiency, against which one may compare the simulated behavior. The idea underlying the efficiency measures is that the four perceived costs of search do not necessarily reflect all the costs of the search process. For example, a change in perceptions concerning the costs of search could lead to different search strategies in which the costs incurred by realty agents are reduced. Such lowered costs could result in agent response that effectively reduces the perceived costs relating to agents (c_3, c_4).

It is clearly possible to construct any number of efficiency measures. In the present study, two measures are chosen for their simplicity. The first is a measure of temporal efficiency, and is the number of searches with agents that are required to locate an acceptable vacancy. The second measure concerns spatial efficiency and is defined as the ratio of the number of searches with an agent in the "good" submarkets to the total number of searches.

For the second part of the analysis it is important to construct adequate, yet simple, measures of the sequential structure of search. We obtain such measures in terms of the one-step transitions from one state to the next, given the action just taken. The analyses are based on the following experimental conditions.[1] Eight subareas are used, only one of which is "good." The location of the good subarea is initially unknown to each decision maker. The set of noncost parameters are

$$\text{PGS (i)} = .125, \text{ PS (jk)} = \begin{bmatrix} 0.4 & 0.1 \\ 0.2 & 0.05 \end{bmatrix}, \text{ R} = 0.8, \text{ PGAR} = 0.2 \text{ and PGAM} = 0.4$$

and characterize all the simulation experiments. Random effects in the simulations enter in relation to obtaining general (C1) information, the outcome of contacting agents (either randomly or by recommendation), and the outcome of search in a given submarket. Standard random number generators produce the random effects. Each simulation experiment continues until the searcher finds an acceptable vacancy or withdraws from the market. At present, we have no results to assess the effects of modifying the experimental conditions. The numbers were chosen for their "reasonableness" and future work is planned to assess such effects.

SEARCH EFFICIENCY AND INFORMATION COSTS

A first set of simulation experiments was run in order to discover the dependence of search efficiency on the relative costs of information. The simulations were run over a full factorial design in the four cost parameters, each of which took the values $c_i = 0.0, 0.1, 0.2, 0.3$, $i = 1,4$. This range of values was chosen so that decision makers always located an acceptable vacancy before withdrawing from the market. Two hundred independent replications were obtained at each cost structure, giving a total of 51,200 observations. The replications were used to compute the *average* number of searches at each cost structure (ANS) and the average number of searches in the good subarea at each cost structure (ANSG). These numbers were also used to compute the ratio of searches in the good subarea to all searches (AR = ANSG/ANS). It should be noted that ANS and AR are, respectively, measures of the temporal and the spatial efficiency of search.

The manner in which ANS and AR depend on the cost structure was investigated by means of regression analysis. The structure of the decision-making model suggests that the two efficiency measures should depend on the four cost parameters in a nonlinear manner. Such dependence was modeled by introducing quadratic and cross-product terms into the regression equations. Since there are two dependent variables, the question of simultaneous equations bias in the parameter estimates arises.

Hence, the regression analysis was conducted in two stages. In the first stage, each dependent variable was regressed against a general second-order polynomial in the costs:

$$ANS_s = \alpha + \sum_{j=1}^{4} \beta_j c_{js} + \sum_{j=1}^{4} \sum_{\substack{k=1 \\ k \geqslant j}}^{4} \gamma_{jk} c_{js} c_{ks} + \epsilon_s; \; s=1,255 \qquad [7]$$

$$AR_s = \theta + \sum_{j=1}^{4} \phi_j c_{js} + \sum_{j=1}^{4} \sum_{\substack{k=1 \\ k \geqslant j}}^{4} \psi_{jk} c_{js} c_{ks} + \omega_s; \; s=1,255 \qquad [8]$$

where α, β_j, γ_{jk}, θ, ϕ_j, ψ_{jk}, are coefficients to be determined and c_{js} are costs. Parameter estimates obtained by using OLS procedures are shown in Table 7.1.

In the second stage of analysis, the correctness of the single equation specification was checked using a 2SLS procedure.[2] Neither dependent variable was found to depend significantly on the other. Hence, it was assumed that the OLS estimates of Table 7.1 provide an adequate approximation to the dependence of ANS and AR on costs.

Variation in the four-cost parameter explains 0.85 of the variance in ANS.[3] There are several features of interest concerning the dependence of ANS on costs, in particular on the costs of general information. First, the average number of searches does not depend significantly on the marginal cost of search with an agent (c_4). Second, an increase in the cost of general information (c_1) causes an increase in ANS, and the dependence is essentially linear. Third, the perceived cost of obtaining a recommendation (c_2) and perceived start-up costs of searching with an agent (c_3) have a significantly nonlinear effect on ANS. For small c_2, c_3 the total number of searches decreases with increasing c_2, c_3, while the rates of decrease themselves decrease with increasing c_2, and c_3, respectively. Search therefore becomes more "efficient" as the costs involved in obtaining recommendations and contacting a new agent increase (at least in the ranges of costs examined). Finally, three of the cost interaction terms have significant coefficients, indicating a relatively complex response of ANS to cost. In particular, since the coefficients of the terms $c_1 * c_2$ and $c_1 * c_3$ are positive, one may interpret the coefficients to indicate an increasing sensitivity of search length to increases in c_2 and c_3 as c_1 increases. The cost of general information thus has a significant effect on the use of other information sources.

The significantly nonlinear response of ANS to information source costs suggests that it may be of interest to determine the cost structure that minimizes ANS (or maximizes the temporal efficiency of search). Using the model whose parameters are specified in Table 7.1, the cost structure that minimizes ANS within the range of cost values considered in the simulations is:

$c_1 = 0.0$,

$c_2 = 0.078$, and

$c_3 = 0.175$.

Hence, the costs of both c_2 and c_3 messages should take on intermediate values. This result is not intuitively obvious, while the effects are not negligible. For example, the cost structure $c_1=c_2=c_3=0.0$ leads to a value of ANS of 3.63, which compares with a minimum value of 3.14. This result emphasizes the importance of inexpensive general information.

The proportion of the variance in the second measure of search efficiency, AR, explained by cost variation is 0.94. The marginal cost of search (c_4) has no affect on this efficiency measure. Costs of general information (c_1) enter in a quadratic manner, while c_2 and c_3 costs enter to a significant degree only in the interaction terms. As c_1 increases, the proportion of searches in the good region decreases for sufficiently small c_1, but at a decreasing rate. As in the previous case, all three interaction terms enter.

Once more it proves of interest to determine the cost structure that maximizes the second measure of search efficiency. The cost values leading to a maximum value of AR are:

$c_1=0.0$,

$c_2=0.3$, and

$c_3=0.0$,

TABLE 7.1 Results of OLS Regressions of Efficiency Measures on Costs

Cost Variable	*Total Searches (ANS)*		*Proportion of Searches in Good Area (AR)*	
	Coefficient	*Probability of t*	*Coefficient*	*Probability of t*
Intercept	3.63	–	.93	–
c_1	2.67	0.01	– .18	0.00
c_2	– 3.05	0.00	– .02	0.88
c_3	– 4.30	0.00	.10	0.34
c_4	– 0.86	0.37	.08	0.47
$c_1{*}c_1$	– 1.32	0.60	1.97	0.00
$c_1{*}c_2$	13.30	0.00	– .83	0.00
$c_1{*}c_3$	14.05	0.00	– .43	0.06
$c_1{*}c_4$	0.70	0.73	.32	0.16
$c_2{*}c_2$	5.85	0.02	.26	0.35
$c_2{*}c_3$	12.16	0.00	– .49	0.03
$c_2{*}c_4$	3.74	0.07	– .14	0.55
$c_3{*}c_3$	9.58	0.00	– .32	0.26
$c_3{*}c_4$	1.17	0.56	– .24	0.29
$c_4{*}c_4$	0.86	0.73	– .21	0.45
	R^2=0.85		R^2=0.93	

which is a boundary point of the region examined. Although this indicates that for maximum spatial efficiency the cost of obtaining a recommendation should be high, the effects are not very marked. For example, the value of AR at $c_1 = c_2 = c_3 = 0.0$ is 0.936, which compares well with the optimum value of 0.945.

In summarizing the results of the regression analysis, it appears that the temporal efficiency of search is, to a large extent, determined by the three costs c_1, c_2, c_3, while the spatial efficiency of search depends largely on c_1. The costs enter in a significantly nonlinear manner, leading to nonintuitive "optimal" cost structures. The marginal cost of search, c_4, has no significant effect on search efficiency. The reason for this lack of effect is that c_4 enters into each of the feasible contemplated action sequences in a uniform manner. Since each sequence contains only one period of search, this cost has no affect on the relative choice of feasible sequences. Finally, it is clear that the cost of general information is the dominant effect governing the efficiency of search.

THE SEQUENTIAL STRUCTURE OF SEARCH

It is hypothesized that the sequential structure of search plays a mediating role between costs and efficiency. Specifically, cost variations affect efficiency because they affect the sequential structure of search, which in turn affects efficiency. A second set of experiments were run in order to examine the influence of cost on the sequential structure of search.

At any point in the search process an individual is characterized by the action just chosen and by a "state." An individual's state may be defined as that set of conditions relating to the individual that determine the next action. One subset of such conditions determines which contemplated action sequences are feasible, and includes (1) whether an individual currently possesses an "unused" recommendation concerning an agent, and (2) whether the individual has already searched with an agent and, if so, whether the agent is good or bad. A second subset of conditions relates to the individual's beliefs concerning whether each of the subareas is good or bad. The conditions relating to beliefs impose problems in determining an individual's state. Since the beliefs are encoded as probabilities, one must define an uncountable set of states. Second, such beliefs are difficult to ascertain in "real" situations. Therefore, we choose to define an individual's state solely in terms of the first set of conditions. The six possible states are

(1) recommendation and good agent located,
(2) recommendation and bad agent located,
(3) recommendation and no agent located,
(4) no recommendation and good agent located,
(5) no recommendation and bad agent located, and
(6) no recommendation and no agent located.

Under this definition of state, an individual's transition from one action to the next is only determined in a probabilistic sense.

Using states and actions, one may define the sequential structure of search as a one-step transition process in two ways First, one may consider the probability of a transition from one action to another, given the individual's state. This approach is discussed in Clark and Smith (1979). Second, one may consider the transitions from one state to the next, given the action that is chosen. We adopt the state transition approach in the present analysis.

There are several reasons for employing the one-step state transition representation of sequential structure. First, one-step transitions provide a simple characterization of sequential structure; yet, the fact that state transitions are action-dependent provides the characterization with some power. In particular, the characterization should prove useful in examining the responses of human decision makers in market and experimental situations and in comparing these responses to model predictions. Second, the representation defines a stochastic finite-state automaton (Paz, 1971) or, equivalently, a production system (see Anderson, 1976; Newall and Simon, 1972). There is a large and growing body of results and techniques for the analysis of such systems; hence, a powerful set of analytical tools is available for future research. Third, the state transition model replaces the extremely cumbersome expected utility model with a much simpler structure. This suggests representing future decision theory models directly in terms of one-step transition models.

In the present study, state transition matrices were constructed from data obtained from simulation experiments. The experiments were run over a full factorial cost structure, in which each of the four costs assumed the values c_i=0.0, 0.5 i=1,4. There were 2000 replications for each of the 16 cost structures.

From the outcomes of the 2000 replications at each cost structure, the number of transitions from each state to every other state were counted. This process was carried out for each of the five actions $a_1 \ldots a_5$ that are involved in state transitions. The result is five 6 x 6 transition matrices for each of the 16 cost structures. A simple, single graphical portrayal of the five matrices for each cost structure may be obtained by plotting the six states as the nodes of a graph and joining any two nodes if they correspond to a nonzero entry in one of the transition matrices. The edges of the graphs are labeled by the action involved in the state transitions. The 16 cost structures induce nine distinct state transition graphs (Figure 7.1). The diagrams immediately reveal some general points concerning the sequential structure of search. First, state one (a message and a good known agent) is never entered. Second, an individual never possesses more than one "unused" recommendation about an agent at any one time. (It should be emphasized that such effects were not built into the simulations as constraints.)

A second set of inductions reveals some general effects of costs on sequential structure. First, it is again evident that the marginal cost of search (c_4) is completely ineffective in modifying the sequential structure of search. Second, when the cost of obtaining a recommendation is low (c_2=0.0), at least one recommendation is always obtained, while none are obtained with a high cost (c_2=0.5). Third, when the cost of obtaining a general message is high (c_1=0.5), general information is only obtained after the searcher locates a good realty agent. Fourth, when the start-up costs of dealing with an agent are low (c_3=0.0), the searcher never searches with a given bad agent more than once.

The preceding results are strong in the sense that there were no exceptions to them in any of the simulation experiments. The graphical portrayal of sequential structure leads easily to strong qualitative conclusions that may be tested in situations involving human subjects. The results also indicate the strong and rather complex effects that subjective information costs have on the sequential structure of search, as in the case of the use of general information in relation to perceptions concerning an agent's ability. These effects presumably underlie the non-intuitive results concerning the cost structures that lead to maximally efficient search.

The model from which the preceding results were obtained is an obvious simplification of the market process. First, there are additional sources of information, such as viewing open houses and driving around. Second, realty agents are more passive in this model than in the real world. Third, the model does not incorporate the possibility of learning new behavior strategies in the market. There are also the restrictions on the range of costs examined and the belief and reliability parameters in these first research efforts. However, even with these simplifications, the model provides useful insights into the sequential structure and efficiency of search. In particular, the results indicate the important role of the costs of general information in structuring the search process and in determining the efficiency of search.

IMPLICATIONS FOR POLICY

Despite specific efforts to reduce constraints on housing choice, there has been little discernible impact on moving behavior. For example, the provision of assistance in the housing allowance demand experiments had no discernible effect on the success of a household search (Weinberg, et al., 1977: A-112) and no significant impact on the probability of a household undertaking search. On the other hand, high transaction costs and the indivisibility of housing services may induce households to tolerate dissatisfaction or disequilibrium until some threshold is reached. Households generally base their housing expenditure on expected or permanent income and not on temporary additions to income. The enforcement of anti-discrimination legislation and the development of legislation to overcome discrimination based on age, source of income, and presence of children may be the most powerful force in affecting search and search costs. Most investigations, however, have not focused on the role of information constraints (or costs) in the housing choice process.

The implications of the simulations reported here emphasize how general information availability can influence the length of search and the resources required for search. This statement is worthy of some elaborative comments in the context of policy. The comments can be organized around notions of availability, disclosure, and discriminatory behaviors on the part of information services.

If we characterize the process of search as involving (a) the length of search, (b) the number of neighborhoods searched, (c)

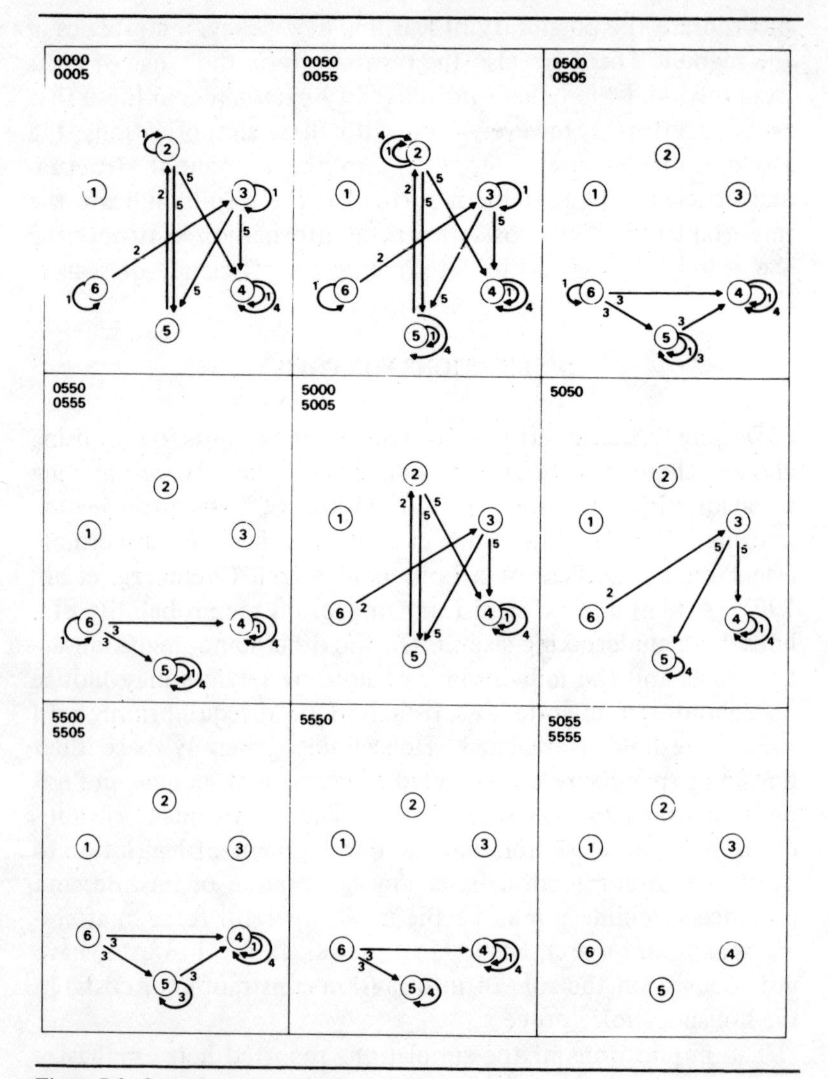

Figure 7.1 State Transition Diagrams

NOTE: Circled numbers indicate states and numbered arrows represent actions. Values for the costs are indicated for c_1 c_2 c_3 c_4 in the upper left of each diagram.

the number of houses examined, and (d) the radius of area searched, we find that all of the search activities are dependent on the number and type of information sources which are utilized and, in particular, on general information. In our model, which only reflects part of the real-world situation, general information is identified with newspaper information, which for most searchers is a main source for an overview of the market. We have already established from the simulations that the use of general information depends on its cost. At the present time the collection of general information from newspapers is not cheap (or reliable, which induces higher costs of usage). The lack of complete disclosure of general market information which is held by real estate brokers in the form of MLS data precludes an accurate overview of the market. Recent litigation (*People* v. *San Diego Board of Realtors* and *People* v. *the San Francisco Board of Realtors*) has focused on access to MLS data. As yet the courts have not decided on the legality of access to MLS data, although the position of the California Association of Realtors is that the information in the MLS data is listed pursuant to a listing contract between an individual seller and a broker. Certainly the issue of disclosure will continue to be of concern to both the public and the real estate industry. The impact of disclosure is even less clear, although it is possible that full disclosure may lead to more efficient search, less time spent with brokers, and more sales accomplished by the most effective brokers.

Another extension of aspects of disclosure relates specifically to the spatial impacts of information. If the evidence of previous studies of residential search and the results of a recent simulation study (Clark and Smith, 1979) indicate that households tend to confine their search to a limited area and that prior knowledge and personal contact based on friends and relatives are among the most important factors in the choice of the area (Speare et al., 1975: 237), then public action could be applied to extend the household's area of familiarity. This relates specifically to disclosure of information. Changing the format and availability of information provision may, in the long run, have a significant impact on the search patterns and final housing choices of mobile households.

The role of discriminatory behavior with respect to information is much more difficult to assess. However, a recently announced study of the residential real estate industry by the Federal Trade Commission intends to investigate (among other topics) "information exchange systems which have the effect of stabilizing prices and restricting competition," "unfairly restricting access to multiple listing services," and "the use of unfair or deceptive forms." Whether or not the widespread six percent selling fee will be adjusted as a result of these investigations is less significant than the increase in search efficiency from better availability and use of information.

CONCLUDING REMARKS

This chapter has argued that information and information availability are critical determinants of the search process. The simulations suggest that information cost is an important dimension in search, and at the present moment general information (as obtained from the reading of newspapers, for example) is inefficient and costly. The means to overcome these problems and affect behavior may only require minor policy changes for perhaps significant results.

NOTES

1. These are the same conditions used to generate the results in Clark and Smith (1979).
2. The choice of which cost variables to include in the two equation specification was made on the basis of the output of single-equation forward stepwise procedures.
3. The unexplained variance can be due only to the random inputs in each simulation run and an inadequate representation of the nonlinearities.

REFERENCES

ANDERSON, J. R. (1976) Language, Memory and Thought. New York: Lawrence Erlbaum.

BARRETT, F. (1973) Residential Search Behavior. Geographical Monograph 1. Toronto: York University.

BERNING, C.A.K. and J. JACOBY (1974) "Patterns of information acquisition in new product purchases." Journal of Consumer Research 1: 18-22.

BETTMAN, J., N. CAPON, R. LUTZ, G. BELCH, and M. BURKE (1978) Affirmative Disclosure in Home Purchasing. Occasional Paper Number 14. Los Angeles: Graduate School of Management, University of California.

CLARK, W.A.V. and T. R. SMITH (1979) "Modelling information in a spatial context." Annals of the Association of American Geographers 69: 575-588.

CRONIN, J. (1979a) An Economic Analysis of Interaurban Search and Mobility Using Alternative Benefit Measures. Washington, DC: The Urban Institute.

--- (1979b) Low-Income Households' Search for Housing: Preliminary Findings on Racial Differences. Washington, DC: The Urban Institute.

GOODMAN, J. (1978) Urban Residential Mobility: Places, People and Policy. Washington, DC: The Urban Institute.

HANUSHEK, E. A. and J. QUIGLEY (1978) "Housing market disequilibrium and residential mobility," pp. 51-98 in W.A.V. Clark and E. G. Moore (eds.) Population Mobility and Residential Change. Studies in Geography 25. Evanston, IL: Northwestern University.

HEMPEL, D. J. (1970) A Comparative Study of the Homebuying Process in Two Connecticut Housing Markets. Real Estate Report Number 10. Storrs: School of Business Administration, University of Connecticut.

--- and McEWEN, W. J. (1975) "The impact of mobility and social integration on information seeking," pp. 361-367 in M. J. Schlinger (ed.) Advances in Consumer Research 2. Proceedings of the Fifth Annual Conference of the Association of Consumer Research.

HERBERT, D. J. (1973) "The residential mobility process: some empirical observations." Area 5: 44-48.

KANARICK, A. F., J. M. HUNTINGTON, and J. M. PETERSON (1969) "Multisource information acquisition and human information processing." Human Factors 11: 403-419.

LEVINE, J. M. and M. G. SAMET (1973) "Information seeking with multiple sources of conflicting and unreliable information." Human Factors 15: 403-419.

--- and R. E. BRAHLEK (1975) "Information seeking with limitations on available information resources." Human Factors 17: 502-513.

McCARTHY, K. F. (1979) "Housing search and residential mobility." Presented at the Conference on the Housing Choices of Low Income Families, Washington, D.C.

MICHELSON, W. (1977) Environmental Choice, Human Behavior and Residential Satisfaction. New York: Oxford Univerity Press.

MORRIS, W. T. (1968) Management Science: A Bayesian Introduction. Englewood Cliffs, NJ: Prentice-Hall.

NEWALL, A. and SIMON, H. (1972) Human Problem Solving. Englewood Cliffs, NJ: Prentice Hall.

PAZ, A. (1971) Introduction to Probabilistic Automata. New York: Academic Press.

ROSSI, P. (1955) Why Families Move. New York: Free Press.

SMITH, T. R., W.A.V. CLARK, J. O. HUFF, and P. SHAPIRO (1979) "A decision-making and search model for intraurban migration." Geographical Analysis 11: 1-22.

SPEARE, A., S. GOLDSTEIN, and W. FREY (1975) Residential Mobility, Migration, and Metropolitan Change. Cambridge, MA: Ballinger.

WEINBERG, D. H., J. FRIEDMAN, and S. K. MAYO (1979) Disequilibrium Model of Housing Search and Residential Mobility. Cambridge, MA: Abt Associates.

WEINBERG, D., R. ATKINSON, A. VIDAL, J. WALLACE, and G. WEISBROD (1977) Housing Allowance Demand Experiment, Locational Choice, Part 1, Search and Mobility. Cambridge, MA: Abt Associates.

8

Contemporary Housing Markets and Neighborhood Change

JAMES T. LITTLE

□ AS WE APPROACH the end of the decade, it is tempting to assert that the fundamental nature of the urban dynamic has been radically altered in many of the metropolitan areas which were "problem" cities in 1970. While the ravaged neighborhoods still exist, even on a massive scale, there is visible reinvestment in both central business districts (CBDs) and residential neighborhoods. Indeed, to rehabilitate a central-city house has become downright fashionable. Nor does rapid neighborhood change, abandonment, and population loss seem to be occurring on the scale of the late sixties and early seventies.

Given the changes in the national economy, a slowing or even a reversal of central-city decline might be expected. The relative price of transportation has risen, increasing the value of access. The supply price of new housing has risen faster than prices in general, thereby increasing the attractiveness of the existing stock. The rapid rate of household formation echoing the post-war baby boom should be increasing demand. Finally, the public programs focused on the central city, ranging from community development spending to antiredlining legislation, might be beginning to have an impact.

AUTHOR'S NOTE: *Financial support for this work was provided by the U.S. Department of Housing and Urban Development under contract number H-2875-G. The conclusions do not reflect the views of HUD.*

At the same time, there is contrary evidence that cannot be ignored. Employment in most of the mature cities has decentralized and continues to do so. In fact, the most rapid rates of job growth are found in fringe counties which are beyond commuting range of the central city. Furthermore, despite the rapid rise in construction costs relative to other costs and the fact that interest rates are high by historical standards, new construction has achieved record levels, with most of the new houses built in newer suburbs. Offsetting the high rate of household formation is the continued population loss of older SMSAs. Thus, while the central city appears to have stabilized, the dynamic of suburban growth seems little changed.

One plausible explanation for this apparent contradiction is that the same process which so visibly affected the central city is still at work but now in the older suburbs. There are several reasons why neighborhood change might be less obvious. That density tends to be lower in the suburbs than in the central city and that the housing is predominantly for single families implies that change will be less concentrated. Since the stock is newer even in the inner suburbs, depreciation of the stock may be less obvious. Because ten years have passed since the last census and because smaller suburban jurisdictions tend to be somewhat unsophisticated in terms of population measurement and reporting, suburban change may be statistically invisible.

The general purpose underlying the empirical analysis developed in this chapter is to assess the extent to which the neighborhood change process continues to operate. Here neighborhood filtering is defined as a change over time of the ranking of neighborhoods according to the location rents associated with them. We construct an indirect test in which filtering is identified and then attempt to relate this test to socioeconomic change within neighborhoods. However, hard statistical evidence of the latter for the post-1970 period is not available; thus, much of the power of the test rests on the similarity of the pattern of neighborhood filtering in the post-1970 period to that identified in earlier work (Leven et al., 1976; Little, 1976).

While the demographic data are not as rich, the coverage both in terms of neighborhoods and time is significantly broader than in earlier analyses of changing neighborhood location rents. This broader coverage allows us to address several subsidiary issues. First, earlier work suggests that socioeconomic change in an adjacent neighborhood has a significant impact on location rents (Bailey, 1966; Berry, 1976). This can be tested as a general proposition through examining the correlations between location rents over time for a large number of neighborhoods. The data also will allow some inferences as to the pattern of site rent changes that are associated with redevelopment of neighborhoods.

THE TRANSITION PROCESS

At the most basic level, neighborhood transition involves a change both in land use and land users. In its declining phase, a change in users occurs from higher-income households to households of lower income, a change in use which frequently involves the conversion of single-family homes to multifamily units. In addition, a general downgrading of the physical quality of the housing stock is experienced, along with an increase in neighborhood population, especially in those cases in which the economic transition is accompanied by racial transition. In its gentrification phase, transition involves replacement of lower-income households by upper-income families and upgrading of the stock.

The economic forces which produce neighborhood succession are best understood when housing is viewed not as a single homogeneous commodity, but rather as a bundle of attributes having physical, locational, neighborhood, and public sector dimensions. In purchasing or renting a dwelling unit, a household gains the right to use the physical structure and land parcel (subject to constraints imposed by zoning and building codes), the right to consume the goods and services produced by the local government, and the responsibility for associated tax liabilities. However, in addition to the legal rights, the household receives a bundle of attributes that can be described as site-specific. The most familiar of these site-specific attributes (at least to one who reads the land use modeling literature) is access to employment and commercial centers. However, any locational attribute that allowed a household to discriminate between alternative locations in terms of preference would also belong to the class of site-specific or "externality" attributes. Thus, if households have well-defined tastes for the socioeconomic status of their neighbors, such neighborhood characteristics as income level and racial mix of the surrounding area are dimensions of the bundle of attributes.

Assuming that households have adequate information as to the attributes associated with alternative sites, the market process will result in a set of equilibrium prices which internalize the externality and public sector components of this bundle of housing attributes, in that the relative desirability or undesirability will be reflected in differential prices. Furthermore, if the externality components of housing bundles are fixed, the allocation of households to sites is efficient in the Pareto sense; that is, there is no rearrangement of households that would not make at least one person worse off. However, it is not the case that the externality components are fixed, particularly for those externalities which are associated with the spatial distribution of land users. For example, in a world without zoning, a tavern can locate at any site for which it can outbid competitors. Presuming that the tavern is a source of a negative externality, initially locating a tavern in a developed residential neighborhood may be suboptimal in the sense that current residents would be willing to pay compensation to the tavern owner sufficient to persuade him not to so locate. However, in subsequent rounds of housing market activity, the market will adjust to an equilibrium which is efficient, conditional on the externality source (that is, the tavern) being at the particular site.

The adjustment to the conditional Pareto-efficient land use assignment will, in general, involve price adjustments in response to the new externality source; that is, if all households view the tavern as the source of a negative externality, prices of houses in the neighborhood will fall. Some residents of the neighborhood, those who have a strong negative preference for taverns, will be replaced by households holding less strong preferences. Irrespective of whether a household moves or stays, it experiences a loss; for movers, it takes the form of capital losses resulting from the decline in housing prices, and for stayers the loss is a utility decline.

The process just described provides an example of the way the movement of an externality source can result in neighborhood succession. However, it is unlikely that movements of such externality sources can explain the extent and intensiveness of succession that has occurred in cities in the past two decades. The explanation lies in a much more pervasive neighborhood effect—the preferences of households with respect to the socioeconomic class of their neighbors. Studies of households' preferences yields the following general conclusions (Berry, 1976; Little, 1976):

- Households reveal themselves to be sensitive to economic status of a neighborhood's residents as measured by mean or median income.
- Households are willing to pay a premium for sites at a distance from concentration of nonwhite population and inferentially prefer neighborhoods with a low probability of socioeconomic change.
- Households reveal themselves to be somewhat more sensitive to the racial composition of adjoining neighborhoods than the racial composition of the neighborhood itself. Furthermore, they are more sensitive to income than to racial composition.
- There are substantial premiums placed on units in "desirable" locations as opposed to low-income, low public service neighborhoods.

The evidence that low-income groups are generally viewed as externality sources suggests a model of neighborhood dynamics based on the notion of a mobile externality source. This model[1], the arbitrage model, has four basic assumptions: (1) other things being equal, all households prefer neighborhoods having higher-income residents to those with predominantly lower-income households; (2) whites have a preference for neighborhoods with low nonwhite proportions; and (3) all households are risk-averse in the sense that they have a preference for neighborhoods in which the probability of socioeconomic change is perceived to be low. It is further assumed that households can be classified into groups according to their race and income; this classification in turn defines "submarkets" on the demand side of the overall housing market.

As an illustration of the mechanism underlying the process, consider a case in which there are two household types, high and low income. The bids of these household types for a given unit depend on the socioeconomic status of the neighborhood in which the unit is located. If we assume that socioeconomic status is measured continuously, we can represent such bids for high (H) and low (L) status (Figure 8.1). As illustrated, high-income households are willing to pay higher premiums for high-status neighborhoods than are low-income households, but they also discount low-status neighborhoods by more than low-income households.

Suppose the unit is located in a neighborhood with status level S_1. The high-income bid prevails, and the unit will sell for a price between L_1 and H_1. Now suppose a reduction in the supply of low-income housing occurs as a result, for example, of a slum clearance program. This will produce a shift in the bid curve of low-income households to L^i. The low-income bid will now prevail, and the price will rise above H_1. However, the change in occupancy will affect the neighborhood's socioeconomic status, leading to decline in status to a level such as S_2. This decline in status will lead to downward pressure on prices.

More important, however, is the fact that the decline in status brought about by the change in occupancy may induce changes in occupancy of other units in the neighborhood. Such a situation is illustrated in Figure 8.2. This case would apply to units of higher quality than considered earlier. Even given the increase in low-income demand, the unit would remain in high-income occupancy provided the neighborhood's status level remained at S_1. However, given the decline in status to S_2, the unit will change from high- to low-income occupancy, inducing a further decline in the neighborhood's status and in prices.

The effect of the reduction in supply in the low-income market, then, is the "arbitraging" of units from the high-income market to the low-income market in response to the higher bid price in the latter. However, there is an important secondary effect: The transfer of units to the low-income market through the effect on neighborhood status induces a shift of additional units into the low-income market. Even though the reduction in supply in low-income areas may have raised the price of sites

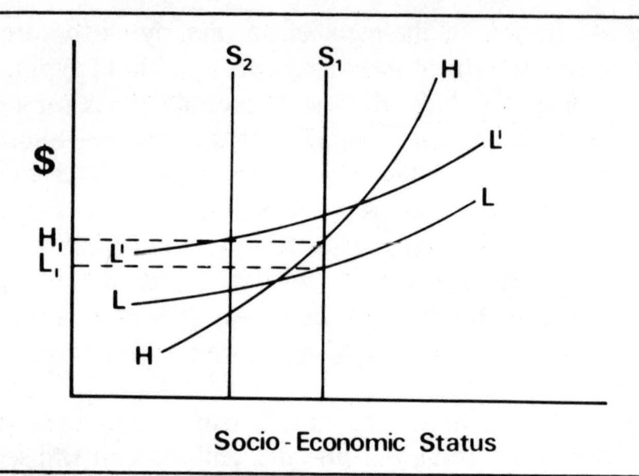

Figure 8.1

generally, there is a possible decline in price (and capital losses for their owners) for these sites which were formerly sold (or rented) in the high-income submarket.

To summarize, the reduction in the supply of sites in the low-income submarket has the following effects: (1) an increase in prices in the low-income submarket; (2) the shifting of units from the high-income submarket into the low-income submarket in response to higher values in the latter; (3) a shifting

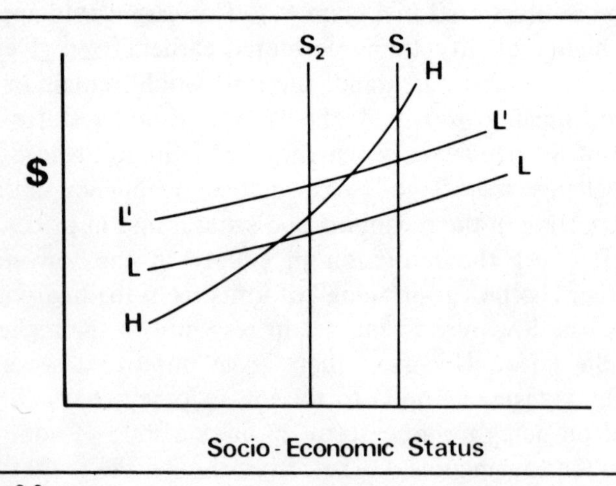

Figure 8.2

of the location of low-income submarkets; (4) a reduction in the price (and of capital values) of sites arbitraged into the low-income submarket; and (5) general upward pressure on prices for sites shifted into the low-income market. Thus, the net effect of the supply reduction is to increase prices except for those units shifted between submarkets.

The description of the adjustment process also applies when the stimulus is an increase in demand in the low-income market or an increase in supply in the high-income market. In both cases, units will shift from the high income to the low-income market. The only difference occurs with an increase in supply in the high-income area. Rather than a general increase in prices, we would expect a decline in prices. Thus, rents decline and capital values decline. However, owners of units which are shifted from the high- to the low- income areas experience an additional decline in capital values resulting from proximity to the low-income area.

The potential for capital losses resulting from changing neighborhood status suggests the role of expectations in the process. Consider the position of an owner-occupant living not in but near the low-income submarket. If his neighborhood were to be absorbed into the low-income submarket and he were to remain in the unit, he will be worse off, since he would be living in proximity to a socioeconomic group which, to him, is less desirable. However, if he were to relocate after transition, he would take a capital loss. The alternative to the less desirable housing bundle or a capital loss is to move deeper into the interior of his submarket's area in advance of transition. But if all potential purchasers of his site have similar expectations as to possibility of transition, the market price of the unit must decline. In Figure 8.3, the shifts of the bid curves to H^i and L^i reflect the increased perceived probability of decline in status. The price of the units will decline, but, more important, if the status of the neighborhood lies between S_1 and S_2, the unit will be shifted from the high-income to the low-income market. The greater the shift in the high-income bid curve relative to that of low-income households, the greater is the probability that a change in expectations will in fact lead to transition. Thus, expectations of transition are self-realizing. Furthermore, once

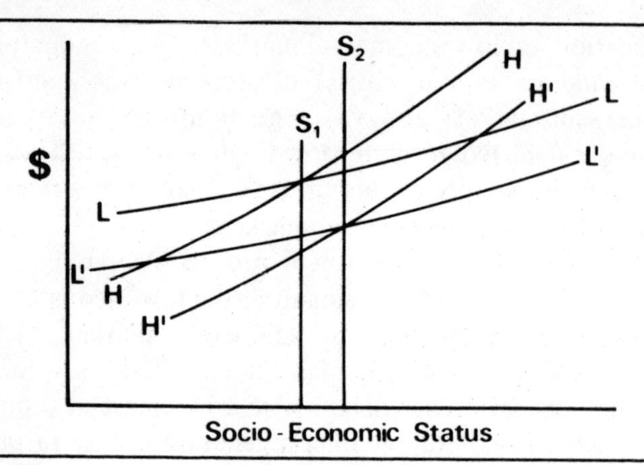

Figure 8.3

transition has occurred, the same process may be repeated for nearby neighborhoods. This suggests that a market characterized by a rapid rate of transition may become dynamically unstable as a result of expectations. That is, even if the supply and demand conditions which produced the initial transition no longer exist, expectations alone may lead to transition.

This instability in neighborhoods abutting transition neighborhoods is compounded by the effect of uncertainty on the lending decisions of financial institutions. In making a real estate loan, the financial institution faces two risks: first, that the borrower will be unable to meet the terms of the loan, and, second, that should default occur, the value of the collateral (the market value of the unit) is uncertain. Given that financial institutions are risk-averse, increased uncertainty as to price or the expectation of lower future prices will result in more stringent lending terms. Since, at some point, these terms will themselves increase the probability of default, increased price risk will ultimately lead to refusal to loan. As price expectations apply to all units in a given area, the refusal to loan manifests itself as "redlining."

The same forces which drive downward transition operate in the gentrification process. For example, the combination of an increase in the number of high-income households (resulting, for example, from a high rate of household formation) and a rising supply curve for new housing will lead to an upward shift of high-income bids. These increased bids on the part of high-income households will lead to the shift of some units from the low-income to the high-income market, which may in turn induce further shifts in response to increases in the status of some neighborhoods. Similarly, when expectations are such as to anticipate an increase in status, there is a tendency toward self-realization.

However, these forces are most likely to produce rising socio-economic status in neighborhoods which are intermediate in status. In neighborhoods dominated by low-income households, the increase in bids by high-income households required to induce gentrification is relatively larger than that required in middle-status neighborhoods. For this reason, significant public investment in neighborhood facilities, particularly in those which appeal to higher-income households, may be necessary if gentrification is to occur.

To summarize, the arbitrage model views the succession process as driven by two forces acting in combination: (1) changing supply and demand conditions in particular submarkets, and (2) expectations as to transition. It is a short-run model in that it takes the physical nature of the housing stock and the general pattern of demand for access as given. However, the fact that in the long run other factors affect neighborhood succession does not negate the importance of the arbitrage model as an explanation of short-run changes. Furthermore, it is the short-run changes that are of greatest significance from a policy standpoint. Perhaps the most important conclusion to emerge from the model is that succession can occur even in the absence of fundamental supply and demand factors. In the presence of high levels of uncertainty and with no market for the trading of these risks (other than that provided by real estate speculators), the market cannot operate efficiently.

LOCATION RENTS AS INDICATORS OF TRANSITION

The basic behavioral force driving the arbitrage model is the generally held preference for "desirable" neighborhoods. As the relative "desirability" of a neighborhood changes, a change in neighborhood population can be expected. Thus, if changes over time in a neighborhood's relative desirability can be measured, these changes provide an indicator of actual or impending change in the neighborhood's population. Location rents provide precisely such a measure of the relative desirability as perceived by participants in the housing market.

Characteristics of housing units can be grouped into broad classes: those which are specific to the unit or parcel and those which depend on location. The first group, the structural characteristics, include such attributes as the size, quality, and style of the structure and the area and topography of the lot. The second group, the neighborhood characteristics, include the socioeconomic character of the neighborhood, accessibility attributes of the location, and tax-expenditure program of the relevant local governmental units. While this list includes attributes which are not normally defined as neighborhood attributes, it includes those characteristics which are beyond the direct control of the owner or occupant of a particular unit.

The proposition basic to the use of location rents as an index of the desirability of neighborhoods is the following: Provided households have well-defined preferences over neighborhood attributes, then if a unit having identical structural attributes sells for different prices in different neighborhoods, the neighborhood having the higher price is preferred to that having the lower price.[2]

The way in which this proposition will be applied using observed price data is straightforward. The price data will be used to estimate the price of a structurally standard unit in a series of neighborhood settings; the preference ranking will then coincide with the ordering in terms of prices of the standard unit. In other words, if the locational rent differential is defined to be the difference in price of these structurally standard units across neighborhoods, then a neighborhood having a positive locational rent differential relative to another ranks higher in the neighborhood hierarchy.

The locational rent differential has certain properties of relevance to this discussion. Only the sign, not the magnitude, of the differential has significance; the sign of the differential is independent of supply conditions, whereas the magnitude of the differential depends on supply conditions. It is this independence of the sign of the differential from supply side conditions which permits inferences as to the changing rank of a particular neighborhood over time. If the price of the structurally standard unit in neighborhood A exceeds the price of standard unit in neighborhood B in 1970, while the reverse is true in 1978, we can conclude that changes in the neighborhood characteristics of A and/or B have occurred which have reversed their relative desirability. Changes in supply conditions and in the characteristics of other neighborhoods in no way affect this inference. On the other hand, no inference can be drawn from the fact that a locational rent differential declines but does not change sign.

To summarize, for any pair of neighborhoods a reversal of the sign of the locational rent differential implies that the characteristics of either or both neighborhoods have changed so as to reverse the relative desirability of the neighborhood. These changes in characteristics may take the form of changes in locational attributes or in the nature of the local public sector. Except in the very long run, however, changes in a neighborhood's socioeconomic character trigger changes in the neighborhood ranking. This fact permits the tracking of neighborhood change through observation of locational rents.

NEIGHBORHOOD CHANGE IN ST. LOUIS

THE SETTING AND THE DATA

To the extent that any city is typical, St. Louis is representative of the "mature" SMSAs, as was its experience in terms of neighborhood change during the 1960s. This experience has been documented extensively elsewhere, but one aspect of this experience should be noted here. By 1970, the most rapidly changing neighborhoods were not in the central city or even in the city itself; rather, they were in the inner tier of suburbs. This provides one clue that neighborhood change in the seventies may be a suburban phenomenon.

The analysis of locational rent changes focuses on 45 "neighborhoods" in St. Louis City and County. These are not "neighborhoods" in the sense that a sociologist would use the term, in that they contain a minimum of two census tracts and typically four. The agglomeration of census tracts was based on several factors, including similarity in terms of access to major employment centers and vintage of the housing stock. Furthermore, tracts were aggregated only if they were located in the same school district and the same city for the larger cities. In St. Louis City, the aggregations coincided with Health District boundaries, which are also used by the city as planning districts. Taken together, the 45 neighborhoods contain about two-thirds of the SMSA population of 2.3 million and cover an extremely wide range of neighborhood types. The time period of the analysis is 1963 to 1978. Not only does this include the period of most rapid change in St. Louis City but also the major structural changes in the economy noted at the beginning of this chapter.

The basic data set consists of selling prices and structural descriptors of over 50,000 single-family houses sold over the 17 years. Using the Census Bureau's DIME file, each parcel was assigned to a census tract and thereby to one of the 45 neighborhoods. A series of socioeconomic, access, and public sector descriptors drawn from a variety of sources were added to the observations for the years 1969 through 1971.

CONSTRUCTION OF THE LOCATION RENT INDEX

As noted above, the preference ranking of neighborhoods will coincide with the ranking across neighborhoods of the prices of a structurally standard unit. The first step, then, in constructing the neighborhood rankings is to define the standard unit. Since this unit must be priced in all neighborhoods, this standardization is not as straightforward as it would first appear. Indeed, given the diversity of the housing stock across neighborhoods in terms of age, style, and quality, it is unlikely that an appropriate unit exists. As a result, the standard unit must be created statistically.

The key to this procedure is the assumed separability of preferences into structural and neighborhood components and the fact that prices rank units according to common preferences. These properties imply that there exist functions f and g, such that the price of a unit can be expressed as

$$p_i = f(\underline{x}_s) + g(\underline{x}_n),$$

where $\underline{x}_s$ and $\underline{x}_n$ are, respectively, the vectors of structural and neighborhood characteristics of unit i. Thus, given f and a vector of standard structural characteristics $\underline{x}_s^*$, the standardized price $\hat{p}_i$ of unit i is given by:

$$\hat{p}_i = p_i - f(\underline{x}_s) + f(\underline{x}_s^*).$$

Using this procedure, the problem of standardization reduces to the problem of estimating f. The function f, however, is not independent of the supply side; consequently, there is no reason to believe that this "valuation" function will remain stable over time.[3] However, this creates another problem, this time one of data. If f is to be estimated via regression techniques, then the appropriate method would be to estimate f separately for each year (assuming the supply side does not change significantly during that period). However, a massive multicollinearity problem exists: Structural characteristics are highly correlated with some neighborhood characteristics, and multivariate estimation techniques appropriate to this context require data on neighborhood characteristics as well as structural characteristics. Since the only source of reasonably accurrate data on socioeconomic characteristics on a small area basis is the census, and, in a world of rapidly changing neighborhoods, socioeconomic characteristics of a given neighborhood can change rapidly, adequate data are simply unavailable.

The approach taken here was to estimate f using the prices of units sold in 1970 together with their structural and neighborhood characteristics. The variables employed in this analysis are listed in Table 8.1. The estimating technique employed was principal components regression, which generated a vector of coefficients $\hat{\beta}$ on structural characteristics.[4] Similar regressions were run on the 1969 and 1971 data sets, and, since the differences between the coefficients estimated using the 1970 data and those based on the 1969 and 1971 samples were not

TABLE 8.1 Housing Characteristics Employed

Price Variables
Market Price (a)
Real Property Taxes (d)
Structural Characteristics
Total usable area of unit (a)
Number of rooms (a)
Bedrooms (a)
Baths (a)
Total area of parcel (a)
Full basement dummy (a)
Central air-conditioning dummy (a)
Age of structure (a)
Socioeconomic Characteristics
Percentage of total population non-white (b)
Highest percentage non-white adjacent tract (b)
Percent all households with female head (b)
Percent all employed population not employed as professionals or supervisors (b)
Median income of families and unrelated individuals (b)
Median school years completed (b)
Percent population less than 20 (b)
Percent all units vacant year round (b)
Percent units constructed before 1939 (b)
Public Sector Characteristics
Expenditure per pupil (c)
Pupil-teacher ratio (c)
Crime rate-crimes per thousand population (e, f)
Access Measures
Peak house travel time to nearest major employment center
Non-peak travel time to nearest major commercial center

SOURCES: (a) Real Estate Transactions File; (b) Census of Population (1970) and Census of Housing (1970); (c) School District Annual Reports; (d) Accessors' Municipalities, St. Louis County Department of Planning; (f) St. Louis Police Commission.

statistically significant, there is some reason to believe that the coefficients, and therefore the valuation function on structural characteristics, is relatively stable over time.

Prices were standardized as follows: First, all prices were converted to 1970 equivalents by deflating them by a 1970-based version of the Bureau of Labor Statistics Home Purchase Index for the St. Louis SMSA. The standard unit was defined as that unit having the mean structural characteristics of the 1970 sample. Prices were then adjusted to reflect the value of the difference in structural characteristics from those of the standard unit. That is, denoting the selling price of the ith unit in the tth year, as p_i^t the standardized price $\hat{p}_i$ is computed as follows:

$$\hat{p}_i = p_i^t \cdot d^t + \Sigma_j \hat{\beta}_j (x_j - \bar{x}_j),$$

where d^t is the deflation factor, $\hat{\beta}_j$ is the estimated valuation coefficient for the jth structural characteristic, x_j is the value of the jth characteristic, and $\bar{x}_j^*$ is the standard value. The neighborhood index was then calculated as the mean standardized price of the units sold in the neighborhood during a given year divided by the mean standardized price of all units sold in that year.

CHANGES IN NEIGHBORHOOD RANKINGS

The main path of socioeconomic change in St. Louis has been along a northwest path from the central business district. By 1962, Health Districts 11 and 20 to the west of the CBD were at that stage of the cycle at which massive abandonment was taking place (see Figure 8.4). To the north and west in Health Districts 9, 10, and 12 lower-income blacks were replacing middle-income whites at a rapid rate. Still farther to the west, Health Districts 5, 6, and 7 were experiencing a moderate but increasing rate of black inmigration.

For the most part, the neighborhoods on the south side were relatively stable, with the exception of Health Districts 23 and 24 immediately to the south of the CBD. These two areas were in many respects similar to Health Districts 9 and 10, in that the housing was of low quality, the residents were poor, and abandonment was occurring at a rapid rate. The difference was that it was poor black households which were leaving the northside neighborhood. Although the remaining neighborhoods on the south side were relatively stable, their populations had been gradually declining and aging.

To the west of the City in St. Louis County (St. Louis City is not part of St. Louis County), the only areas which were "changing" were the suburban growth areas: Hazelwood to the north, Parkway North and South to the west, and Lindbergh and Mehlville to the south.[5]

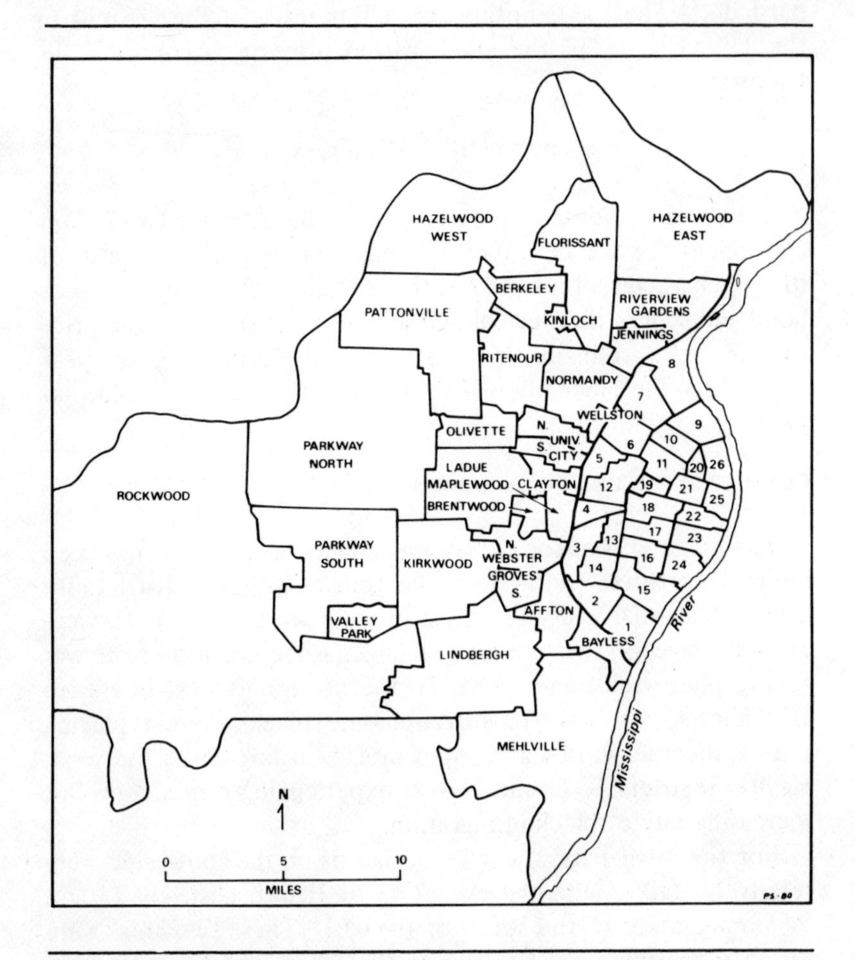

Figure 8.4 **Health Districts in the City of St. Louis and School Districts in the County of St. Louis**

By 1970, the situation had changed dramatically. Health Districts 5, 6, 9, 10, and 12 in the city had reached the abandonment phase of the cycle, and rapid change had moved northwest into Health District 7 and had spilled west across the city-county boundary into Wellston and University City North and South. Further north, Normandy was beginning to experience some change. Rapid suburban development was occurring in Parkway South and Rockwood, but the most rapid rates of suburban growth were now to be found still farther west outside St. Louis County.

Given this context, let us now turn to the analysis of the neighborhood rankings generated by the location rent index. These rankings for the years 1963, 1966, 1970, 1974, and 1978 are given in Table 8.2; the ranks run from 1 to 45, where 1 is the highest rank and 45 the lowest.

Several general observations can be made. The first is that there has been no general increase in the ranking of city neighborhoods relative to those in the county. In 1963, 13 of the 20 lowest-ranked neighborhoods were in the city; in 1970, 17 of 20; and in 1978, 16 of the lowest 20. On the basis of this evidence, there appears to have been little change in the perceived attractiveness of city neighborhoods relative to other alternatives. Nor does there appear to be a discernible pattern which would indicate a decline in the relative attractiveness of neighborhoods on the suburban "fringe." Of these growth areas, Rockwood and Lindbergh have risen significantly in rank since 1963, while the two Parkway neighborhoods have maintained a stable and high ranking. Pattonville is the only fringe neighborhood to decline significantly in rank between 1963 and 1978, but the biggest change came between 1963 and 1966, not after 1970. Furthermore, while three relatively centrally located neighborhoods in the county—Kirkwood and the two Webster Groves areas—did experience a significant increase in rank, the increase occurred, for the most part, before the increase in gasoline prices.

The most important general observation that can be made, however, is that this evidence strongly suggests that the same dynamic which was at work in the city in the 1960s is now playing itself out in the suburbs. The pattern of location rents over time predicted by the arbitrage model and observed in typical changing neighborhoods is the following: Because change is anticipated, location rents tend to decline before change occurs; if change is initiated by supply decrease in low-income areas, prices and rank may temporarily rise. However, as the lower-income group begins to dominate the neighborhood population, location rents and the relative ranking of the neighborhood declines. The rankings over time of Health District 7 and of University City North are typical of this pattern: As the group of lower economic status began to move into the neighborhood around 1966, the ranking rose, but as

TABLE 8.2 Neighborhood Rankings

St. Louis County	*1963*	*1966*	*1970*	*1974*	*1978*
Affton	16	11	16	25	15
Berkeley	33	26	27	27	28
Clayton	7	2	2	2	4
Florissant	1	8	20	14	20
Hazelwood East	19	12	13	10	19
Jennings	24	23	25	24	31
Kirkwood	14	13	7	7	7
Ladue	5	1	1	1	1
Lindbergh	12	7	6	8	6
Maplewood	26	25	26	31	21
Mehlville	3	6	11	5	9
Normandy	31	19	21	29	30
Olivette	10	3	8	17	12
Parkway North	8	5	3	3	3
Parkway South	4	14	4	4	2
Pattonville	2	15	14	9	14
Ritenour	29	22	18	28	23
Riverview	9	10	19	19	24
Rockwood	20	24	5	16	10
University City North	27	20	23	23	25
University City South	17	4	9	26	11
Webster Groves North	25	21	17	20	18
Webster Groves South	15	9	10	6	5
Wellston	39	35	34	22	32

St. Louis City	*1963*	*1966*	*1970*	*1974*	*1978*
Health District 1	28	32	24	11	27
2	13	16	15	12	13
3	22	27	22	33	17
4	34	29	31	36	22
5	43	40	39	45	34
6	37	37	35	39	39
7	35	28	29	32	35
8	30	18	28	18	30
10	40	36	33	35	37
11	44	42	40	44	44
12	6	39	44	38	8
13	11	33	37	15	29
14	18	17	12	21	16
15	21	31	30	30	33
16	36	34	32	34	36
17	42	41	43	41	40
18	38	44	38	43	42
19	23	*	45	13	43
20	*	30	41	37	38
23	32	43	42	42	45
24	41	38	36	40	41

*Insufficient sales to calculate reliable index number.

the neighborhood is shifted into the lower-income market, the ranking again declines.

Given the general northwesterly movement of change through the city, the county areas in which succession was likely to have occurred in the 1970s are Normandy, Jennings, and University City South; and, if the process is operating quickly, Florissant, Hazelwood East, and Riverview are possible candidates. For five of the six neighborhoods, the exception being University City South, the trends in the rankings are consistent with the predictions of the arbitrage model. In Normandy, for example, we see an increase in rank in 1966 and then a sharp decline between 1970 and 1978. At the same time, the proportion of blacks in the Normandy school system rose from about 30 percent in 1970 to almost 80 percent in 1978. While these school enrollment statistics certainly overstate the change, they are indicative of its magnitude. While the pattern for Jennings is not so clear (for there is no sharp increase in rank that would reflect the beginnings of the inmigration of the lower-income population), the significant decline in rank is consistent with declining socioeconomic status of neighborhood residents. The decline in rank of Riverview foreshadows the change that is now becoming apparent in school enrollments. Similarly, the declining rank for Hazelwood East, which lies north of Riverview, reflects anticipated change.

The sharpest pattern of all is that for Florissant. The largest decline in rank occurred between 1963 and 1970. There are several factors underlying this decline. First, Florissant is adjacent to the first urban expressway to be completed in the St. Louis area, and, as the expressway system was extended during the late sixties and early seventies, the northern tier of county neighborhoods lost their relative advantage in terms of access. A second factor was the increased likelihood of change as Normandy and Jennings experienced socioeconomic succession. Finally, it surely also reflects that, under a 1975 court order, the Ferguson-Florissant School District was forced to merge with that of Kinloch, a "black" town which had existed as a stable, although low-income, enclave for more than 70 years. While the most visible indications of neighborhood change are not yet evident in Florissant, both the decline in neighborhood rank and the rapid turnover of the housing stock which is currently taking place indicate that it is likely—although not certain—that significant change in economic status will occur.

The experience of University City South demonstrates that massive population turnover and declining status are not the inevitable consequences of the arbitrage process. While this neighborhood did experience some modest change in its racial composition and income status during the late sixties and early seventies, there has been a sharp reversal in the direction of location rent changes since 1974. There is no single factor that can explain this reversal. The housing stock is of high quality, but this was also true for neighborhoods which had not experienced a reversal in decline. Although University City had adopted a point-of-turnover housing code enforcement system and occupancy controls in 1968, similar systems in municipalities within Normandy did not appear to either slow turnover rates or increase maintenance. However, these factors, together with the rising supply of housing and University City's location in the center of the central corridor of the St. Louis metropolitan area—a corridor in which the major centers of professional employment are located—were probably important contributory factors. And while it is not clear whether the hypothesis can be tested, the magnitude of the reversal may be simply a result of the fact that several years of stability indicates to both current residents and potential buyers that decline is not inevitable.

University City South represents a neighborhood in which the change process was reversed: It is not, however, a neighborhood that has come back from rock bottom. Within the group of neighborhoods, however, there is one which appears to qualify: Health District 12, known to St. Louisians as the Central West End. If we ignore the 1963 ranking, the increase in the rank of this area is simply remarkable. After ten years of a very low ranking, between 1974 and 1978, it rises 30 ranks. The increase in the relative attractiveness of the neighborhood is brought into focus even more sharply by the fact that apartment buildings which ten years ago were sold for back taxes have become $100,000 condominiums. Furthermore, there are numerous examples of doubling and tripling of single-family house prices in the past five years.

Like University City, the Central West End is located in the central corridor, which would be appealing to the managers and professionals who now tend to occupy the neighborhood. Another factor in the improvement has been the high levels of community development spending in the area and a large-scale redevelopment project of Washington University Medical School. However, without doubt, the most important element in the neighborhoods recent success was its former high-status position. The neighborhood's relatively high ranking for 1963 reflects the fact that the Central West End had been the most elegant neighborhood in the city and, indeed, in the whole metropolitan area. Even up to 1970, the commercial district was a center for high-status retailers. Furthermore, if University City's housing stock is of high quality, that of the Central West End is unique: The single-family units are primarily Victorian townhouses or turn-of-the-century mansions. The multifamily buildings are of comparable quality. Thus, if one considers the longer-run view, the redevelopment of the Central West End neighborhood is really a return to its former rank. The success should in no way be depreciated by that fact, particularly given that redevelopment has proved to be a difficult process in St. Louis. However, at this time, we cannot escape the inference that success depended critically on the neighborhood's unique character.

CONCLUSIONS

If any strong conclusion emerges from the analysis, it is that the rather dramatic changes in the economic environment which occurred in the 1970s have had little effect on the attractiveness of central-city neighborhoods relative to suburban areas. This conclusion in no sense diminishes the success of redevelopment efforts; indeed, it makes such success even more impressive. It does indicate, however, that the same forces which led to outmigration from central cities in the fifties and sixties are still at work, and that, while rates may slow, net migration from the central cities to the suburbs can be expected to persist.

This is not to say that nothing has changed. It is now clear that the process of socioeconomic transition, once begun, can be aborted. The experiences of University City South and, to some extent, Olivette provide evidence of this. What is important in the experience of these two areas is that the process stabilized and even reversed itself without massive public intervention. In fact, it can be argued that the experiences of these two communities demonstrate the symmetry of the arbitrage process. Just as supply and demand factors can set a process of downward transition in motion, it can be reversed; in this case, by the rising supply price of new housing. However important the rising supply price may have been to neighborhoods, the evidence suggests it was insufficient to reverse the transition process on a widespread basis.

The above observation brings us to the conclusion that is—although tentative—the most important from the point of view of public policy: the transition process is still operating on a significant scale, but is becoming a suburban rather than a central-city phenomenon. This conclusion has important implications for public policy because urban policy tends to focus on places rather than problems. Thus, while the stress on central cities in our current programs may have been appropriate in the sixties and seventies, it may be inappropriate in the eighties.

The policy response that will be required goes beyond a broadening of the focus of existing programs. By virtue of their populations, central cities retain a comparatively sophisticated bureaucracy. Many of the suburban jurisdictions have virtually no professional staff, and, even if they have one, it is generally too small to design and implement programs in response to transition. More important, smaller suburban jurisdictions rarely have the ability to marshall economic resources on a significant scale. A central city, for example, by concentrating its Community Development Block Grant funds in a few neighborhoods, can have an immediate and significant impact. Since small suburban municipalities are rarely entitlement cities, their CDBG funds will typically be received as part of an allocation to a county. Given the politics of allocating these funds to competing municipalities, it is unlikely that the focusing of block grant funds, possible in the central city, is politically feasible in the context of suburban government.

Clearly, these problems would not exist if suburban government were reasonably centralized. However, it is equally clear that there is little possibility of such centralization occurring in the near future. In fact, the very existence of transition in some communities and not in others makes the possibility of cooperation even more remote. The design of urban programs, then, must recognize that transition and the difficulties of adapting to it are not exclusively central-city problems and that programs must be adaptable to fundamentally different governmental contexts than that of the central city.

NOTES

1. This model is largely based on the work of Bailey (1959) and is fully elaborated in Leven et al. (1976).
2. For a detailed discussion see Little (1975).
3. We can interpret f as a hedonic functional over structural characteristics. A discussion of the role of the supply side in determining this functional can be found in Rosen (1974).
4. Details of the estimating technique and the estimates of the coefficients can be found in Little (1976) and Mark and Parks (1970). The $\hat{\beta}$'s employed are those of column 3 of Table 1 in the latter paper.
5. Wellston and University City North are exceptions. These neighborhoods are adjacent to those areas of the city in which the rate of socioeconomic changes was escalating.

REFERENCES

BAILEY, M. J. (1966) "Effects of race and other demographic factors on the values of single-family homes." Land Economics 42: 215-220.

--- (1959) "A note on the economics of residential zoning urban renewal." Land Economics 35: 288-292.

BERRY, B.J.L. (1976) "Ghetto expansion and single-family house prices: Chicago, 1968-72." Journal of Urban Economics 3: 397-423.

KAIN, J. F. and J. M. QUIGLEY (1975) Housing Markets and Racial Discrimination. New York: National Bureau of Economic Research.

LEVEN, C. L., J. T. LITTLE, H. O. NOURSE, and R. B. READ (1976) Neighborhood Change. New York: Praeger.

LITTLE, J. T. (1976) "Residential preferences, neighborhood filtering and neighborhood change." Journal of Urban Economics 3: 68-81.

--- (1975) "Rents and preferences in assignment housing markets." Working Paper HMS 5. St. Louis: Institute for Urban and Regional Studies, Washington University.

MARK, J. and R. P. PARKS (1970) "Neighborhood preferences, neighborhood filtering and neighborhood change: comment and corrections." Journal of Urban Economics 5: 535-537.

ROSEN, S. (1974) "Hedonic prices and implicit markets: product differentiation in pure competition." Journal of Political Economy 82: 34-55.

Part III

Modeling the Impacts of Public Programs

□ THE PAST DECADE has witnessed the proliferation of several major experimental programs which impact on housing and mobility. Among these are the housing allowance and the income maintenance programs (see Quigley and Hanushek, 1979, for a discussion of their impacts). They are among the first programmatic attempts to influence the housing behavior of the population. Within the framework of these experiments, a variety of specific modeling strategies have been outlined and evaluated. The third section of this book is devoted to a general discussion of the role of modeling in mobility research and to two specific examples drawn from the housing allowance experience.

In an introductory and general chapter, Alan Wilson sets out to examine models of residential location and their responsiveness to the concerns of public policy. He suggests that traditional models of mobility have been of only peripheral interest to policy-making, partly because policy variables have not been included explicitly in the models. Moreover, the variety of problems which face policy makers are never made clear to the research community. The ensuing chapters by Daniel Weinberg and Mark David Menchik are very specific applications of modeling strategies to data generated in the housing allowance experiments. Menchik develops a new analytical procedure based on the concept of a hazard function which can be applied to the type of censored data on mobility derivable from administrative records. This procedure is used to examine the lengths of stay of households in order to understand the effects of housing allowances, with different eligibility rules, on the likelihood of movement in the supply experiment. Weinberg reviews the mobility behavior exhibited by test and control groups in the demand experiment and analyzes responses in relation to different experimental treatments. He sets the research in the context of the process by which households adjust their housing consumption to desired or equilibrium levels, with mobility constituting the prime mechanism for the adjustment.

These studies raise a number of issues concerning the utility of modeling strategies from the perspective of different public policy clients. At the federal level, to which both Weinberg and Menchik direct their chapters, a major emphasis is placed on generalizability even though the experimental programs are city-specific. Concern is for the general effects of such programs, effects which are sufficiently strong that they might be expected in a wide variety of local contexts. Experimental design is vitally important, and, in that regard, Peter Rossi pointed out at the conference that the demand experiment has the substantial advantage of a well-defined control group of nonrecipients which is absent in the supply experiment. However, the size of the program is such that the outcomes still tend to be dominated by the "normal operations of the local market"; in addition, the characteristics of the local markets tend to differ quite sharply, leading, for example, to much higher mobility rates in Phoenix and Pittsburgh (the two sites in the demand experiment). The combination of these observations implies that if substantive results are to be generalized, much more effort needs to be given to inclusion of local contextual variables within the structure of these models.

Once we shift from the federal to the local level, further questions arise about the usefulness of models of this type. Both Peter Morrison and Ralph Ginsberg pointed out during the conference that policy makers at the local level are often not in a position to make highly structured decisions. They are placed in a problem-solving context in which solutions to complex issues must be presented within highly constrained time frames. Most proposed solutions are tentative at best and must bear close political scrutiny if they are to survive. Under these circumstances, the basics of the modeling effort and the meaning of the output must be clear, concise, and easily understood. Mathematical or economic models will often fail this crucial test. In fact, Ginsberg argued that we need to pay much more attention to the ways in which various organizations in the policy field cope with information which is presented at different levels of sophistication.

This is not to say that models are not of value; but those which are of use to policy makers are likely to be simpler and more robust than those presently developed in the context of academic research. They must be structured in such a way that interpretation of the role of those instrumental variables available to policy makers is possible. At the local level, the need for elementary information about types of household, dwelling unit, and neighborhood change has brought a resurgence of interest in accounting models and the types of data necessary to support them. Such models are often simple and readily understood and communicate effectively. Simulation models can substitute for expensive data collection and still provide insights into alternative manipulations of instrumental variables. Perhaps the dominant idea to emerge from these chapters was that academics do not pay enough attention to the context in which their work is to be used. To contribute to the policy field, results and the basic argument backing those results must be both relevant to specific policy issues and clear to decision-makers.

REFERENCE

QUIGLEY, J. and E. HANUSHEK (1979) Complex Public Subsidies and Complex Household Behavior: Consumption Aspects of Housing Allowances. Working Paper 825. New Haven, CT: Institute for Social and Policy Studies, Yale University.

9

Residential Mobility: Policy, Models, and Information

ALAN G. WILSON

□ THE OBJECTIVE OF THIS CHAPTER is to review models of residential location and change in the context of their information needs and their contributions to public policy. Given the breadth of the topic, only a broad review is feasible, and a somewhat polemical stance is adopted. In the next section, the policy context of this style of modeling is reviewed briefly; two sections on modeling follow. In the first of these, the general characteristics of models in the residential field are outlined briefly, and then a number of topics are discussed relating to rate estimation, the task of modeling mobility processes, and the measurement of policy impacts. A number of preliminary conclusions are drawn in these sections; these are brought together more systematically in a concluding section, along with a discussion of strategies for future work.

THE POLICY CONTEXT

It is often stated that models of residential mobility have contributed little to public policy discussions (Clark and Moore, 1978; Moore and Harris, 1979). It is important, therefore, to begin with a brief analysis of why this should be the case. There

AUTHOR'S NOTE: *I am grateful to Huw Williams for helpful discussions in relation to this paper though I bear the responsibility for its contents and particularly its shortcomings. A longer version of the paper containing a more detailed discussion of models is available as Working Paper 251, School of Geography, University of Leeds.*

are probably two main reasons. First, the models do not contain variables which seem relevant to policy makers. It will be argued below that this is a situation which can be rectified as far as the models are concerned. Second, the public policy issues in the field are poorly articulated in spite of an enormous existing literature. On both counts, therefore, we need to examine how the policy context can be clarified.

Planning and policy questions arise from a variety of perspectives. From first principles, we would begin with people and households and their lifestyles. Planning can be seen as concerned with seeking ways to improve their packages of activities and opportunities (housing, transport, services, employment, and so on). This involves surveys and predictions in relation to these topics and the identification of "problems." Planning and policy are then concerned with the resolution of these problems, or, more positively, with the achievement of various goals. There are always fundamental difficulties of procedure with this viewpoint: While it is often easy to identify problems and solutions to problems, the resources are not usually available to implement solutions. There is competition for the available resources, and their allocation can be seen as the outcome of class conflict. The nature of the competition and conflict varies across different countries and societies but is usually easily recognizable. The argument can be extended from individual and household needs to community or social needs, and this recognizes the existence of groups of people who form pressure groups or whose interests may be represented by others—for example, by planners. The same issues of competition and conflict will be present.

Planning and policy organizations formulate their own goals: They see themselves as representing particular groups in particular instances and possibly in trying to resolve conflicts. These organizations may be, in the broadest sense, resource-based (a housing agency may be taken as an example of this category) or place-based. For completeness, we should also note the policy ambitions and roles of a wide range of organizations whose functions are economic; these may be in both private and public sectors.

Broadly, then, we can argue that there are two types of policy questions: those concerned with people or communities from first principles–the identification of problems and what can be done about them in a context of competition for resources and of class conflict–and those of particular, usually governmental, organizations which ask: "Given our specific brief, how can we improve and implement our policies?" There is also a third, superpolicy question: "Are the present institutions adequate for handling current and future problems?"

When this background is related to housing and, more specifically, to residential mobility, some uncomfortable observations immediately can be made. First, it is clear that in all large cities, for example, and in most other situations as well, many people have housing problems perhaps associated with jobs, services, and access to opportunities. One of the reasons why policy is not well articulated is that this would involve the admission that the resources are not available to tackle most of these problems in the foreseeable future. Thus, the first step, it can be argued, is to recognize this and to concentrate some effort in spelling out the boundaries of what is achievable.

The interdependence of people's problems is also a difficulty. Many mistakes have been made in housing policy because housing problems have been treated in isolation. This is something which must not be forgotten in deciding what kinds of models offer the best contributions to policy.

The variety of agencies involved causes difficulties. Any one agency usually offers only a particular contribution to the resolution of housing problems. If these efforts could be coordinated, and the systemic consequences of particular actions recognized, then better use could be made of the resources which are available. It is beyond the scope of this essay to consider any particular policy agencies. The next step in the argument, therefore, is to discuss in principle what the instruments of policy are in order to provide a background for the discussion of modeling and information needs; because of the level of generality adopted, it is hoped that any conclusions can be translated easily into the context of a particular organization.

There are, broadly, four kinds of policy instrument: (1) expenditure, (2) regulation, (3) fiscal policy, and (4) form of organization. All are relevant to the housing and residential mobility field. The difficulties to be borne in mind at present include the current aversion to increased levels of public expenditure, especially at times of economic recession, and the vested interests which have accumulated for the beneficiaries in relation to past policies on regulation, taxation, and forms of organization. A few examples will suffice to sustain the argument: Land-use controls, or the lack of them, favor landowners in a particular way; tax policies in Britain provide tax relief for interest on mortgages and generate enormous benefits, especially in inflationary times, for the owner-occupying class; many people in particular agencies have an interest in the survival of these agencies in terms of their own jobs and careers.

Nonetheless, the policy variables fall under one or more of these four headings. These are the variables which, once explicitly presented, we should seek to represent in our models. We review the adequacy of this, sometimes directly and sometimes by implication, in the outline of models in a subsequent section. We return to this issue specifically in the concluding section.

GENERAL CHARACTERISTICS OF MODELS OF RESIDENTIAL LOCATION AND CHANGE

Any residential model is part of a broader model system. This is important in at least two respects. First, people's choice of residence will be partly determined by variables from other submodels, such as those representing the supply of services or income. Such variables will have to be either modeled themselves or treated exogenously. Second, the same point becomes particularly significant in another sense if the residential model is to be used in a policy context. This will involve at least the measurement of short-run response to policy and possibly longer-term projections. In such a case, it is more difficult simply to deem other variables to be exogenous.

A residential model has a number of main components: people and households by type; houses by type; an allocation of households to houses; and perhaps some related determinants of behavior, such as a representation of the residential environment. This provides a static picture of residential location and, obviously, the basis of a sequence of comparative static pictures. The next step is to add dynamics and process; this involves the organizations responsible for all aspects of housing supply and related features of residential choice. The "types" adopted in the system description should reflect the different kinds of behavior thought to represent change. The enormous number of categories apparently necessary for this forms one of the main problems of model construction.

The various entities can be defined at different scales: this is partly a matter of choosing the breadth of classes which make up types; partly a matter of the phenomena involved–decisions by individuals may be influenced by macro variables such as quality of the environment, which may in turn be seen as an aggregate of many individual decisions; and partly a matter of model design and data availability. Typically, we would then seek an accounting framework: The basic demographic processes and housing supply processes form a basis for the study of change and a framework for keeping track of all possible and actual transitions. Finally, we would seek a theoretical base for any model: an understanding of the processes of change and the way in which these ideas can be built into a model.

Given this background, what kinds of models would we expect to find? It has often been useful to use Weaver's (1958) classification of system and problem types as an aid to exploring what kinds of models are likely to be appropriate to particular situations (Wilson, 1977). He identified three types: (I) simple systems, described by only three or four variables; (II) systems of disorganized complexity, described by thousands or millions of variables, but with only weak interactions among the entities of the system; and (III) systems of organized complexity, which also involves large numbers of variables but with some strong interactions among them. The first type is of no relevance here, except possibly at a very aggregate scale (for example, for housing supply as part of an econometric model). We might expect the allocation of households to houses to be represented by a Weaver-II model, since the number of cooperative effects is small and in any case these can be built explicitly into the models. This suggests that statistical averaging models will be appropriate; these could be either of the average-rate accounting type (for example, in the Markov family) or of the entropy-maximizing kind. It can also be argued that various methods used for building models involve the same broad kind of averaging to solve the aggregation problem—whether these be the probabilistic methods of random utility theory or the methods of mathematical programming. It is not surprising, therefore, that we find all these methods represented in our survey of models in the next section. When the supply side is added, the system probably becomes Weaver-III. Two explanations account for this: First, there are strong interactions among many of the agencies involved; and, second, in many cases the numbers of agencies (for example, those involved in house-building) will not be sufficiently large to make any sort of averaging feasible. There are no general methods available for building models of such systems, but one interesting method for dynamic model-building which seems relevant to this context, but which has not yet been applied, is referenced below.

MODELS

AN OUTLINE OF MODEL TYPES

The factors or variables which have been included in models of mobility arise from a great deal of empirical work and are the basis of much verbal theorizing. The range of model "types" is also large; a classification is outlined below, with example references, to serve as a context for an evaluation of their usefulness in the policy avenue.

(1) Linear models focused on mobility rates (Moore, 1969).
(2) Markov models (McGinnis, 1968; see Ginsberg, 1973, 1978, for examples of "rate" models; see Moore, 1978; Byler and Gale, 1978, for examples of accounting models).
(3) Search behavior models (Smith et al., 1979).
(4) Economic models (Alonso, 1964; Muth, 1969; Mills, 1972; Goodman, 1976; Hanushek and Quigley, 1978).

(5) Dispersion models (entropy-maximizing or random utility; Senior and Wilson, 1974; Wilson and Senior, 1974; Williams, 1977).
(6) Optimization models (Coelho et al, 1978; Coelho, 1977).
(7) Simulation models (Kain et al., 1976; Wilson and Pownall, 1976; Clarke et al., 1979).
(8) Dynamical modeling (Harris and Wilson, 1978; Wilson, 1979).

In the rest of this section we present a brief discussion of aspects of the use of these models under various headings. We distinguish rate estimation models from models of process, and then discuss the derivation of policy impacts.

RATE ESTIMATIONS

In modeling, the residential location decision is usually broken down into "whether" and "where to." This parallels the distinction between trip generation and distribution in transport modeling; suggesting that the main use of linear models is likely to be in the estimation of total rates of movement to or from an area, and that regression analysis is one possible technique. However, it is surprising that the method known as category analysis (Wootton and Pick, 1967), which has been used extensively in trip generation analysis, has not been used in the residential mobility context. It could provide a short-run robust working tool. If its use were coupled with appropriate monitoring procedures, then such estimation of rates could be updated as appropriate. (There are, of course, formal similarities between category analysis and regression analysis with dummy variables, or with contingency table analysis.)

It is clear that more fundamental questions are involved in rate estimation—analogous to seeking the reasons underlying birth processes in demographic analysis—and we pick up the threads of this topic in the discussion in the next subsection on modeling processes.

While it seems a reasonable approximation to model total rates with linear models, we implied above that the answers to the "where" question could not reasonably be modeled in this simple way. There are, however, analogous rate-based models—the Markov models. Again, these may provide useful short-run models and an accounting foundation which can ultimately be applied in more sophisticated models. This process involves the appropriate definitions of time periods, states, and transitions (see Rees and Wilson, 1977, for an application of such ideas in spatial population analysis at a coarser spatial scale). At present, Markov models as applied to residential mobility take too little account of space in state definitions, but this could easily be rectified.

MODELS OF PROCESSES

The processes underlying residential mobility can be formulated in various ways and incorporated into models in different ways. The broadest framework is probably that offered by economic modeling, based on the theory of consumers' behavior (utility-maximizing) on the demand side and the theory of the firm (probably as profit-maximizing) on the supply side. Most recent work is based on the seminal studies of such authors as Alonso, Muth, and Mills. The most developed side, and on the whole the most relevant for studies of residential mobility, relates to demand. The utility function contains a representation of housing preferences, usually in the form of bid rents. The main achievements of the early urban economists was perhaps to show that a comprehensive model could be built which included a market-clearing mechanism and that an equilibrium solution existed. Goodman (1976) and Hanushek and Quigley (1978) criticized the equilibrium assumptions of this kind of model and showed how to compute the divergence from equilibrium of particular households—the "drift" from equilibrium over time. They related this to the costs of moving, which are neglected in many models of residential mobility, and showed how the two concepts can be used as the basis for a probabilistic model of residential relocation.

A number of problems can be identified where further advances are needed. First, economic models often imply "rational and perfect" behavior in a world where responses are more "dispersed." It is interesting, in this respect, to compare the economic models with those based on entropy-maximizing methods, which might be described as essentially dispersion models (see Wilson, 1970, for a broad account and Senior,

1973, for a detailed review of the residential location context—which precedes his review of economic models: Senior, 1974). These models have proved robust in a number of fields and are designed to make further developments relatively straightforward. It is easy, for example, to incorporate new terms into utility functions or their equivalent (which are the "attractiveness" and transport terms of interaction models) and to calibrate and test such models using either aggregate or individual household data. It can also be argued that they have a better "behavioral" base than may have been thought. They lend themselves to the inclusion of additional constraints and also to "information"-adding procedures, which will be discussed again in a broader context below (Batty and March, 1976). This class of models has an advantage, in that there is a wide variety to choose from to meet many tastes in theory. These models can be used in conjunction with other models, such as the Markov models as noted earlier; although it should also be emphasized that they often need inputs from other models, such as a mobility rates model. In particular, they can be combined with economic modeling ideas, so that dispersion can be introduced in the economic context. Senior and Wilson (1974) combine approaches from Alonso (1964) and Herbert and Stevens (1960) on the one hand with entropy-maximizing on the other.

A second deficiency of economic models is in their treatment of the supply side. Again, there is the possibility of a new approach to the problem of dynamic modeling which has its origins in spatial interaction models but which is applicable more widely (Harris and Wilson, 1978; Wilson, 1979). Models will emerge which will contain representations of new phenomena based on concepts of critical parameter values. This will have an obvious importance in planning. The difficulty will lie in the calibration and testing of these models: The data demands—involving time series—are likely to be more stringent than in any known models. So while the theoretical insights may be important in the short run, this is likely to be a longer-run development for applied work.

Third, it could be argued that insufficient attention has been paid to search aspects of the residential mobility process. However, this may not be the most fruitful way of proceding, especially if not informed by empirical research. There may be an argument for more empirical work which may then lead to model-building. However, there are two ways in which search behavior can be reflected in existing models: one implicit and one explicit. We saw above that dispersion can be included in many models, and one feature of this could be a representation of the uncertainties and imperfections of search behavior. Second, if it is thought that search behavior biases model results which do not incorporate it explicitly, then it could be built into spatial interaction models as some kind of "search cost"—to be added to the usual cost terms, but including directional biases or whatever empirical research was found to be appropriate.

IMPACTS OF POLICIES

The models listed earlier and reviewed in different contexts in the section above can all be used in obvious ways to measure the impact of policies. The key issue, noted previously, is the way in which the policy variables are incorporated into the models—usually as elements of supply, such as houses or jobs of particular types at particular locations, or as costs or benefits to consumers. Models must be made sensitive to what are seen as the main policy variables in particular contexts. The direct impacts of policies can be measured by the response of the exogenous variables of the models: what sort of people live where, at what time, both before and after a change. It is also possible to compute indices such as changes in consumers' surplus relative to the cost of implementing a policy, thus estimating social rates of return. The systemwide impacts of policy can be traced through other submodels: when the endogenous variables of the residential model form inputs to other submodels.

For both direct and systemwide impacts, it is appropriate to draw attention to optimization and simulation models which were listed earlier but not discussed specifically. Their main roles are at the policy impact stage. First, if it is clear what the policy variables are, and if goals can be explicitly articulated, then it is possible to apply formal optimization methods. The mathematical programming formats involved then both provide direct evidence on effectiveness through the objective function–for example, maximization of consumers' surplus or minimization of costs–and, possibly more important, indirect information through dual variables. Second, micro-simulation models may be applied, particularly in estimating systemwide impacts; it is difficult to retain all information simultaneously in conventional algebraic model formats, but the simulation method offers a way of overcoming this (Wilson and Pownall, 1976).

CONCLUDING COMMENTS

MODEL AVAILABILITY

It is clear from this brief survey that there is a wide range of models available which contributes to the study of residential mobility; these form a rich basis for future work. A number of suggestions have been made about how these can be combined in various ways and also how the field may have much to learn from modeling developments in adjacent fields in urban studies. There are some new problems which are just beginning to be effectively tackled–such as the study of the effects of "chaining"–and other fields where it may be a long time before effective applied work is possible–for example, in the study of criticality. Nonetheless, the substantive conclusion is that a good range of models is available for both analytical and policy work.

MODELS AND DATA AVAILABILITY

There is no doubt that if an ideal model were designed, it would outstrip data availability. However, one of the main arguments of this chapter has been that models can be adapted to make the best use of available data. In effect, knowledge can be accumulated in a Bayesian way, although this may demand a particular form of organization of work (discussed briefly below). It is probably also true that the British record is better than the American one in this respect. There are a number of examples of the development of large-scale models for which some of the data are missing, but submodels have been added or various estimation principles adopted to fill in the gaps in such a way as to allow model-building to go ahead (see Rees and Wilson, 1977, for examples in the demographic field and Wilson et al., 1977, for a wide range of examples). In general, much use can be made of standard data, such as that from censuses; these data can be supplemented by a variety of special surveys in particular cases.

POLICY VARIABLES AND MODELS

There is not a good record in the field of building variables into models which are sensitive to policy (Gale, 1978). In part, this involves some model development which will be discussed further below. However, in general, much progress could be made with available models. What is needed is much more experience in the use of models in a policy context. Some additional research may be needed on the formulation of evaluation measures, but this issue is closely connected to the difficult task of policy articulation discussed above. When policy objectives can be clearly articulated, there should be no difficulty, in principle, of constructing appropriate evaluation measures from model outputs.

THE ORGANIZATION OF FUTURE WORK

If models are to make more contributions to policy, there is probably no substitute for the relevant agencies becoming more directly involved. In this way, knowledge can be accumulated in one place; this also makes a direct contribution to the data problem. Monitoring is also important in this context, both for its direct usefulness and its ability to help in data accumulation. There are useful lessons to be learned from the transportation studies here. The policy and planning agencies which have gained most from these studies are those that have made considerable in-house effort which could be maintained and continued after the initial study.

The other point to stress in this context is the importance of variety in the work to be carried out. Housing is part of a larger system, and interdependencies are important both in understanding and in assessing the impacts of policy. This implies that there should be some large-scale model-building efforts on the scale of the NBER work, for example. However, there are also many possible smaller studies which could be carried out. Models could be designed ad hoc in relation to particular problems. It is in light of this kind of experience, for which there is no substitute, that the best decisions can be made about the future of models of mobility which are relevant for policy.

REFERENCES

ALONSO, W. (1964) Location and Land Use. Cambridge, MA: Harvard University Press.

BATTY, M. and L. MARCH (1976) "The method of residues in urban modelling." Environment and Planning, A 8: 189-214.

BYLER, J. W. and S. GALE (1978) "Social accounts and planning for changes in urban housing markets." Environment and Planning, A 10: 247-266.

CLARK, W.A.V. and E. G. MOORE (1978) "Data bases and research issues," pp. 1-12 in W.A.V. Clark and E. G. Moore (eds.) Population Mobility and Residential Change. Studies in Geography 25. Evanston, IL: Northwestern University.

CLARKE, M., P. KEYS, and H.C.W.L. WILLIAMS (1979) "Household dynamics and economic forecasting: a micro simulation approach." Presented at the Regional Science Association, London, August 1979.

COELHO, J. D. (1977) "The use of mathematical optimization methods in model based land use planning. An application to the new town of Santo Andre." Ph.D. thesis. University of Leeds School of Geography.

--- and A. G. WILSON (1976) "The optimum size and location of shopping centres." Regional Studies 10: 413-421.

COELHO, J. D., H.Ç.W.L. WILLIAMS, and A. G. WILSON (1978) "Entropy maximizing submodels within overall mathematical programming frameworks: a correction." Geographical Analysis 10: 195-201.

GALE, S. (1978) "Remarks on information needs for the study of geographic mobility," pp. 13-14 in W.A.V. Clark and E. G. Moore (eds.) Population Mobility and Residential Change. Studies in Geography 25. Evanston, IL: Northwestern University.

GINSBERG, R. B. (1978) "Probability models of residence histories: analysis of time between moves," pp. 233-265 in W.A.V. Clark and E. G. Moore (eds.) Population Mobility and Residential Change. Studies in Geography 25. Evanston, IL: Northwestern University.

--- (1973) "Stochastic models of residential and geographic mobility for heterogeneous populations." Environment and Planning, A 5: 113-124.

GOODMAN, J. L. (1976) "Housing consumption disequilibrium and local residential mobility." Environment and Planning, A 8: 855-874.

HANUSHEK, E. A. and J. M. QUIGLEY (1978) "Housing market disequilibrium and residential mobility," pp. 51-98 in W.A.V. Clark and E. G. Moore (eds.) Population Mobility and Residential Change. Studies in Geography 25. Evanston, IL: Northwestern University.

HARRIS, B. and A. G. WILSON (1978) "Equilibrium values and dynamics of attractiveness terms in production-constrained spatial-interaction models." Environment and Planning, A 10: 371-388.

HERBERT, D. J. and B. H. STEVENS (1960) "A model for the distribution of residential activity in urban areas." Journal of Regional Science 2: 21-36.

KAIN, J. F., W. C. APGAR, and R. J. GINN (1976) Simulation of the Market Effects of Housing Allowances. Volume 1: Description of the NBER Simulation Model. New York: National Bureau of Economic Research.

McGINNIS, R. (1968) "A stochastic model of social mobility." American Sociological Review 23: 712-722.

MILLS, E. S. (1972) Studies in the Structure of the Urban Economy. Baltimore: MD: John Hopkins University Press.

MOORE, E. G. (1978) "The impact of residential mobility on population characteristics at the neighborhood level," pp. 151-81 in W.A.V. Clark and E. G. Moore (eds) Population Mobility and Residential Change. Studies in Geography 25. Evanston, IL: Northwestern University.

--- (1969) "The structure of intra-urban movement rates: an ecological model." Urban Studies 5: 17-33.

--- and W.A.V. CLARK (1978) "Data structures and anlytic strategies in migration research," pp. 267-279 in W.A.V. Clark and E. G. Moore (eds.) Population Mobility and Residential Change. Studies in Geography 25. Evanston, IL: Northwestern University.

MOORE, E. G. and R. S. HARRIS (1979) "Residential mobility and public policy." Geographical Analysis 11: 175-183.

MUTH, R. F. (1969) Cities and Housing. Chicago: University of Chicago Press.

REES, P. H. and A. G. WILSON (1977) Spatial Population Analysis. London: Edward Arnold.

SENIOR, M. L. (1973) "Approaches to residential location modelling II: urban economic models and some recent developments." Environment and Planning, A 6: 369-409.

SENIOR, M. L. and A. G. WILSON (1974) "Disaggregated residential location models: some tests and further theoretical developments," pp. 141-172 in E. L. Cripps (ed.) Space-Time Concepts in Urban and Regional Models. London: Pion.

SMITH, T. R., W.A.V. CLARK, J. O. HUFF, and P. SHAPIRO (1979) "A decision-making and search model for intra-urban migration." Geographical Analysis 11: 1-22.

WEAVER, W. (1958) "A quarter century in the natural sciences," pp. 7-22 in Annual Report. New York: The Rockefeller Foundation.

WILLIAMS, H.C.W.L. (1977) "On the formation of travel demand models and economic evaluation measures of user benefit." Environment and Planning, A 9: 285-344.

WILSON, A. G. (1979) "Aspects of catastróphe theory and bifurcation theory in regional science." Papers, Regional Science Association 00: 000-000.

--- (1977) "Recent developments in urban and regional modeling towards an articulation of systems theoretical foundations," in Proceedings, Vol. 10 giornate di Lavoro. v AIRO Parma:

--- (1970) Entropy in Urban and Regional Modelling. London: Pion.

--- and C. POWNALL (1976) "A new representation of the urban system for modelling and for the study of micro-level interdependence." Area 8: 246-254.

WILSON, A. G. and M. L. SENIOR (1974) "Some relationships between entropy maximizing models, mathematical programming models and their duals." Journal of Regional Science 14: 207-215.

WILSON, A. G., P. H. REES, and C. M. LEIGH [eds.] (1977) Models of Cities and Regions. Chichester, England: John Wiley.

WOOTTON, H. H. and G. W. PICK (1967) "A model for trips generated by households." Journal of Transport Economics and Policy 1: 137-153.

10

Mobility and Housing Change: The Housing Allowance Demand Experiment

DANIEL H. WEINBERG

□ MOVING is the process by which substantial housing change takes place. Rarely is residential mobility a goal in itself; rather, it is a means to an end—a larger apartment, a new house, and so forth. Yet, often mobility has been studied in isolation, rather than as a process resulting in housing change. The analyses of mobility in the Housing Allowance Demand Experiment illustrate a dual approach: Residential mobility along with household search are examined in isolation (why do families move?), but they are also studied as key ingredients in determining household response to the experimental treatments (what happens when they move?). This chapter is an attempt to draw together these two strands of research as they were carried out in the analysis of the demand experiment data.

AUTHOR'S NOTE: *Financial support for the research on which this chapter is based was provided by the U.S. Department of Housing and Urban Development under Contract H-2040R to Abt Associates Inc. I wish to thank my co-workers, whose research is cited herein, particularly Joseph Friedman, William Hamilton, Stephen Kennedy, Jean MacMillan, Shirley Mansfield, Stephen Mayo, Sally Merrill, and James Wallace. Nancy Burstein and Peter Rossi also provided valuable comments on an earlier draft. Responsibility for this study remains, of course, with the author.*

THE HOUSING ALLOWANCE DEMAND EXPERIMENT[1]

In recent decades American society has been concerned with at least two fundamental problems of housing for poor people: physically inadequate housing and housing costs that place too heavy a burden on meager incomes. Governmental responses to these problems have ranged from constructing and maintaining low-income housing to making cash payments to poor people. Housing allowances fall between the direct income transfer and conventional public housing approaches to solving these problems, some features of which already appear in the recently established Section 8 Existing (Leased) Housing program. The housing allowance is not as locationally restrictive as conventional public housing because recipients choose for themselves how much to spend for housing and where to live, whereas conventional public housing typically involves construction of new units in specific locations. However, it is more constrained than welfare programs because the subsidy depends in some way on the housing chosen. A housing allowance does not impose the kinds of geographical constraints for which public housing has been criticized, but neither does it promote economic or racial deconcentration through the use of specific neighborhood or locational requirements.

The demand experiment was designed to provide a test of how well a housing allowance strategy works. The experiment was one of three conducted by the Department of Housing and Urban Development as part of the Experimental Housing Allowance Program. Operated in Allegheny County, Pennsylvania (Pittsburgh), and Maricopa County, Arizona (Phoenix), the demand experiment offered renter households, selected at random from among the potentially eligible households at each site, one of several housing allowance plans. Households remained in the experimental program for three years after they were enrolled. During their third year, households were offered help in transferring to ongoing housing programs. Analysis of program effects is based on the data for the first two years. Figure 10.1 provides a detailed schematic and sample sizes for the plans described in greater detail below.

HOUSING GAP: (P = C - bY, where C is a multiple of C*)

b VALUE	C LEVEL	HOUSING REQUIREMENTS: Minimum Standards	Minimum Rent Low = 0.7C*	Minimum Rent High = 0.9C*	No Requirement
b = 0.15	C*	Plan 10 PIT = 45 PHX = 36			
b = 0.25	1.2C*	Plan 1 PIT = 33 PHX = 30	Plan 4 PIT = 34 PHX = 24	Plan 7 PIT = 30 PHX = 30	
	C*	Plan 2 PIT = 42 PHX = 35	Plan 5 PIT = 50 PHX = 39	Plan 8 PIT = 44 PHX = 44	Plan 12 PIT = 63 PHX = 40
	0.8C*	Plan 3 PIT = 43 PHX = 39	Plan 6 PIT = 44 PHX = 35	Plan 9 PIT = 43 PHX = 35	
b = 0.35	C*	Plan 11 PIT = 41 PHX = 34			

Total Housing Gap: 512 households in Pittsburgh, 421 households in Phoenix.

Symbols: b = Rate at which the allowance decreases as the income increases.
C* = Basic payment level (varied by family size and also by site).

PERCENT OF RENT (P = aR) :

a = 0.6	a = 0.5	a = 0.4	a = 0.3	a = 0.2
Plan 13 PIT = 28 PHX = 21	Plans 14 - 16 PIT = 109 PHX = 81	Plans 17 - 19 PIT = 113 PHX = 66	Plans 20 - 22 PIT = 92 PHX = 84	Plan 23 PIT = 65 PHX = 46

Total Percent of Rent: 407 households in Pittsburgh, 298 households in Phoenix.

CONTROLS:

With Housing Information	Without Housing Information
Plan 24 PIT = 159 PHX = 137	Plan 25 PIT = 162 PHX = 145

Total Controls: 321 households in Pittsburgh, 282 households in Phoenix.

Figure 10.1 Demand Experiment Housing Allowance Plans and Sample Size after Two Years

NOTE: This sample includes households that were active, although not necessarily receiving payments, after two years of enrollment; households whose enrollment income was above the eligibility limits or which moved into subsidized housing or their own homes were excluded. While data on the excluded households may be useful for special analyses, particular analyses may also require the use of a still more restricted sample than the one shown here.

The basic plan, called "Housing Gap," offered payments large enough to bridge the gap between the cost of modest, existing standard housing and a reasonable fraction of income. As in the Housing Assistance Supply Experiment, the allowance payment was linked to housing by requiring recipients' housing to meet certain requirements. Variations in both payment levels and housing requirements were tested. "Minimum Standards" requirements specified certain physical conditions on recipients' housing, including the presence of bath and kitchen facilities and an adequate number of rooms for the size of the family. The alternative type of requirement, "Minimum Rent," specified only that a household of a given size spend some minimum amount on housing. Families already living in adequate housing (as defined by the program's requirements) were allowed to use the allowance payment to reduce the burden of housing costs. Those not in adequate housing had to improve their current housing or move to a qualifying unit in order to receive the allowance. In either case, households electing to participate were provided resources to obtain decent, affordable housing.

Another experimental plan offered a payment using the same formula as the Housing Gap plans but had no housing requirement. This plan, classified as "Unconstrained," resembles welfare or other general income support programs, except that the subsidy is determined by an expected need for housing expenditures rather than an expected need for all household expenses. An alternative type of housing allowance plan, the "Percent of Rent" approach, offered a rent rebate. Allowance payments were a fixed fraction of monthly rent. Control households, a final group enrolled in the experiment, did not receive a housing allowance but instead received a $10 monthly payment for providing information.

The demand experiment thus provided a controlled experiment in housing strategies and established an empirical basis for assessing a wide range of housing and income transfer policies. It should be emphasized, though, that the findings pertain to the behavior of low-income households.

THE PROCESS OF MOBILITY

Residential mobility in the demand experiment was studied directly by MacMillan (1978). It was precisely because of the central role of mobility in household behavior that an intensive effort was made to understand the process itself. As MacMillan stated:

> Because of the key role of mobility in housing and neighborhood improvement and in program participation, it is important to learn what factors [in addition to] the allowance offer influenced households in their decision to move. If some households were less likely to move than others, then they may have been less likely to participate or to obtain better housing or neighborhoods. It is especially important to know if households did not move because of difficulties that might have been alleviated by program assistance, such as information about the housing market, help with childcare or transportation, or equal opportunity support.
>
> Analysis of mobility during the Demand Experiment concentrated on whether the allowance offer affected mobility, whether some groups of households were less likely to move than others, and whether any program actions might have made it easier for households to move [1978: S-2].

MacMillan's analysis was focused on estimating experimental effects. To that end, she adopted an "eclectic" approach that incorporated the major variables found to be related to mobility in other studies. In order to further understand why demographic and site differences occurred, she broke mobility down into four stages--becoming dissatisfied, planning to move, searching, and moving. I will focus here on only the last stage.

It appears that natural patterns of mobility prevail. Even without an allowance offer, most low-income renters would move over the course of several years. In Pittsburgh, 43 percent of the households enrolled in the experiment had moved at least once in the two years before the experiment began. In Phoenix,

71 percent had moved at least once in those two years. The experiment did result in some increased mobility–there was a significant difference in the mobility rates of experimental and control households of about seven percentage points in the sites combined, consisting of about 10 percentage points in Phoenix, a statistically significant difference, but an insignificant difference of five percentage points in Pittsburgh. Table 10.1 presents the results of fitting a logit model to movement data from the cities.[2]

The particular treatment plan involved played an important role. Percent of Rent households had significant increases in mobility in both sites, while Unconstrained households had significant increases in mobility only in Pittsburgh (see Figure 10.2). In contrast, Housing Gap households as a whole had significant increases in mobility only in Phoenix. This response differed by whether housing requirements were met at enrollment (see Figure 10.3). Significant increases in mobility were apparent for Housing Gap households that did not meet their requirements at enrollment in both sites. In contrast, the experiment appeared to have provided some incentive for Pittsburgh Housing Gap households that met the requirements initially to stay where they were, while having the reverse effect in Phoenix. This site difference is puzzling, particularly as it persists in other analyses.

MacMillan's analysis also covered other factors. The widely reported finding that older households were less likely to move was confirmed: Age had a significant, negative relationship to the probability of moving at both sites. Further, there was some evidence that minority households were less likely to move in Pittsburgh, but not in Phoenix.

THE OUTCOMES OF MOBILITY

Household housing changes in response to the allowance offer were studied in two parts–locational change and housing change. These two areas will be discussed in turn.

TABLE 10.1 Logit Estimation of the Probability of Moving

	Pittsburgh			Phoenix			Combined Sites		
Independent Variable	*Coefficient*	*Asymptotic t-Statistic*	*Partial Derivative*[a]	*Coefficient*	*Asymptotic t-Statistic*	*Partial Derivative*[a]	*Coefficient*	*Asymptotic t-Statistic*	*Partial Derivative*[a]
Mobility History									
Number of moves in three years prior to the experiment	0.253	4.39**	0.058	0.409	7.89**	0.100	0.331	8.97**	0.082
Length of residence in enrollment unit (in years)	−0.051	−4.27	−0.012	−0.029	−1.14	−0.007	−0.043	−3.23**	−0.011
Housing Factors									
Perceived crowding	0.386	2.55*	0.089	0.273	1.87†	0.067	0.338	2.89**	0.084
Living in a unit with basic features	−0.325	−2.68**	−0.075	−0.212	−1.77†	−0.052	−0.267	−2.84**	−0.066
Social Bonds									
Positive feelings toward neighbors	−0.065	−2.08*	−0.015	−0.126	−4.08**	−0.031	−0.087	−4.37**	−0.021
Dissatisfaction									
Dissatisfaction with unit or neighborhood at enrollment	0.423	3.15**	0.097	0.135	0.91	0.033	0.328	3.33**	0.081
Predisposition to Move									
Would move with an increase in money available for rent	0.726	4.92**	0.167	0.604	5.34**	0.148	0.658	8.10**	0.163

TABLE 10.1 (Continued)

Independent Variable	Pittsburgh			Phoenix			Combined Sites		
	Coefficient	Asymptotic t-Statistic	Partial Derivative[a]	Coefficient	Asymptotic t-Statistic	Partial Derivative[a]	Coefficient	Asymptotic t-Statistic	Partial Derivative[a]
Constant	0.022	0.04		0.076	0.69		−0.269	−0.72	
Life Cycle Factors									
Age of household head (in decades)	−0.230	−4.33**	−0.053	−0.239	−6.09**	−0.059	−0.233	−6.17*	−0.058
Number of children	−0.036	−0.76	−0.008	−0.015	−0.35	−0.004	−0.029	−0.81	−0.007
Change in number of children	0.141	1.02	0.032	0.296	1.72†	0.073	0.213	2.04*	0.053
Change in marital status	0.947	3.94**	0.218	0.800	3.22**	0.196	0.869	5.12**	0.215
Other Household Characteristics									
Female head of household	0.301	2.41*	0.069	0.436	3.28**	0.107	0.340	3.66**	0.084
Black head of household	−0.369	−2.61**	−0.085	0.494	1.88†	0.121	−0.155	−1.27	−0.038
Spanish American head of household				0.003	0.02	0.000	−0.131	−0.89	−0.032
Years of education of household head	−0.038	−1.29	−0.009	0.001	0.03	0.000	−0.019	−1.00	−0.005
Per capita income of household (in thousands)	0.126	1.12	0.029	0.102	1.10	0.025	0.119	1.70	0.030

TABLE 10.1 (Continued)

Independent Variable	Pittsburgh			Phoenix			Combined Sites		
	Coefficient	*Asymptotic t-Statistic*	*Partial Derivative*[a]	*Coefficient*	*Asymptotic t-Statistic*	*Partial Derivative*[a]	*Coefficient*	*Asymptotic t-Statistic*	*Partial Derivative*[a]
Program Factors									
Experimental household	0.224	1.56	0.052	0.391	3.07**	0.096	0.297	3.27**	0.073
Site									
Phoenix							0.649	8.47**	0.161
Likelihood Ratio (significance)		209.13**			221.45**			499.39**	
Sample Size		1,037			795			1,832	
Mean of Dependent Variable		0.359			0.572			0.451	
Coefficient of Determination		0.154			0.204			0.198	

[a] The partial derivative represents the change in probability, given a unit change in the independent variable (evaluated at the mean of all independent variables).
† t-statistic significant at the 0.10 level (two-tailed).
* t-statistic significant at the 0.05 level (two-tailed).
** t-statistic significant at the 0.01 level (two-tailed).
SOURCE: MacMillan (1978: Table 3-4) and MacMillan (forthcoming: Table 2).

LOCATIONAL CHANGE

Analysis of locational change in the demand experiment was undertaken by Atkinson et al. (1979). Of all the outcomes considered, this is clearly the most dependent on mobility. No locational change can occur without moving, yet mobility does not imply change because local, intraneighborhood moves occur often.

A key feature of a housing allowance, in contrast to most of the more traditional forms of housing assistance, is that it allows participants substantial freedom in their choice of residential locations. Households offered allowances in the demand experiment could live anywhere in the program area (Allegheny County or Maricopa County), provided that their dwelling units met program requirements. The freedom of locational choice inherent in the housing allowance concept has prompted speculation that the program would lead to large-scale redistribution of the population. The most frequent conjecture was that an allowance would allow the low-income population to disperse to higher-income areas and allow minorities to move into more integrated locations.[3]

As with MacMillan's analysis, the focus of Atkinson et al.'s analysis was to estimate the effect of the experimental programs with little effort directed toward developing a behavioral model of locational choice. The methodology they chose was straightforward: Since random assignment of households across experimental and control programs was used, they compared the patterns of locational choice in the population offered housing allowances with the patterns observed in an equivalent population not offered an opportunity to participate.[4]

The bottom line of their analyses was that no significant change occurred:

> The availability of the housing allowance did not induce households to choose neighborhoods with the significantly different economic compositions than those they would have chosen in the absence of a program [see Table 10.2]. . . .

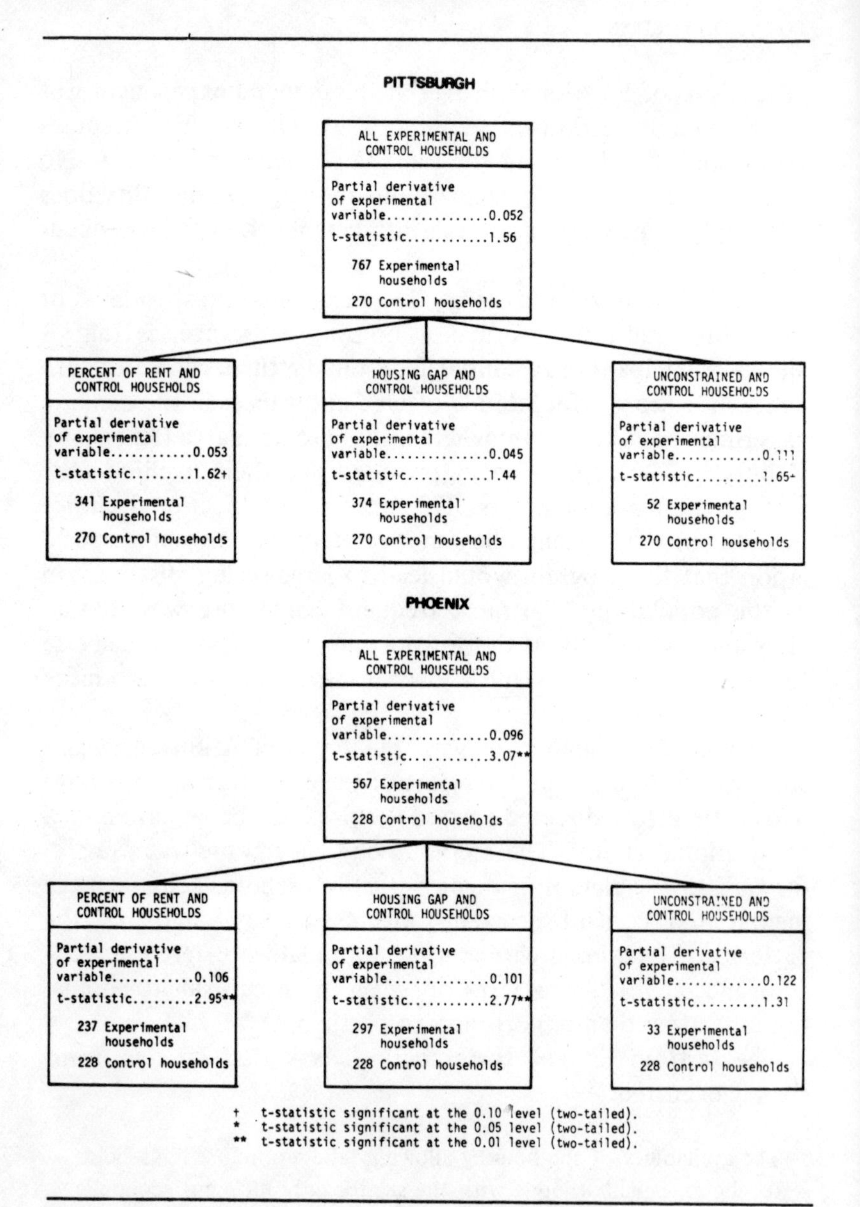

Figure 10.2 Effect of the Experiment on Mobility by Treatment Group
SOURCE: MacMillan (1978: Fig. 4.2).

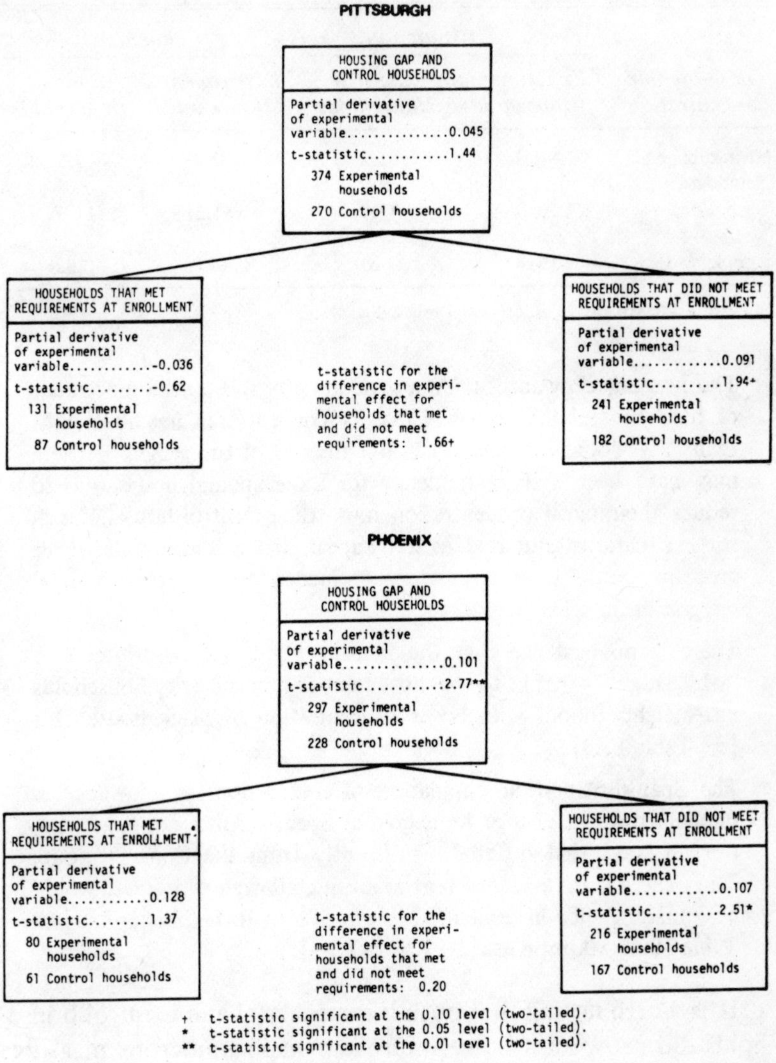

PITTSBURGH

HOUSING GAP AND CONTROL HOUSEHOLDS
Partial derivative of experimental variable..............0.045
t-statistic...........1.44
374 Experimental households
270 Control households

HOUSEHOLDS THAT MET REQUIREMENTS AT ENROLLMENT
Partial derivative of experimental variable...........-0.036
t-statistic.........-0.62
131 Experimental households
87 Control households

t-statistic for the difference in experimental effect for households that met and did not meet requirements: 1.66+

HOUSEHOLDS THAT DID NOT MEET REQUIREMENTS AT ENROLLMENT
Partial derivative of experimental variable............0.091
t-statistic..........1.94+
241 Experimental households
182 Control households

PHOENIX

HOUSING GAP AND CONTROL HOUSEHOLDS
Partial derivative of experimental variable..............0.101
t-statistic...........2.77**
297 Experimental households
228 Control households

HOUSEHOLDS THAT MET REQUIREMENTS AT ENROLLMENT
Partial derivative of experimental variable...........0.128
t-statistic.........1.37
80 Experimental households
61 Control households

t-statistic for the difference in experimental effect for households that met and did not meet requirements: 0.20

HOUSEHOLDS THAT DID NOT MEET REQUIREMENTS AT ENROLLMENT
Partial derivative of experimental variable............0.107
t-statistic..........2.51*
216 Experimental households
167 Control households

+ t-statistic significant at the 0.10 level (two-tailed).
* t-statistic significant at the 0.05 level (two-tailed).
** t-statistic significant at the 0.01 level (two-tailed).

Figure 10.3 Effect of Compliance with Requirements at Enrollment on the Mobility Housing Gap Households

SOURCE: MacMillan (1978: Fig. 4.3).

TABLE 10.2 Mean Change in Low-Income Concentration (percentage points)

	Pittsburgh		*Phoenix*	
Low-Income Concentration	*Experimental Households*	*Control Households*	*Experimental Households*	*Control Households*
Mean change (standard deviation)	−1.1 (8.1)	−1.2 (7.2)	− 2.7 (11.3)	− 3.3 (11.0)
Sample size	(916)	(320)	(715)	(282)

SOURCE: Atkinson et al. (1979: Table 2-3).

> The housing allowance did not generate any substantial movement of black households into less racially concentrated neighborhoods than they would have chosen in the absence of the program. There may have been a slight tendency for Experimental households to reduce their racial concentration more than Control households in some situations. But it does not appear that a housing allowance program would have any strong influence on patterns of racial integration [see Table 10.3]....
>
> There is no evidence that the housing allowance contributed to "white flight"—that is, to the movement of non-minority households into neighborhoods with lower concentration of black households [see Table 10.3]....
>
> The Spanish-American population offered a housing allowance in Phoenix did not change its degree of Spanish-American concentration in ways that differed significantly from the Control group. Thus, there is no evidence that a housing allowance program would be a major factor in residential integration of this ethnic group [see Table 10.4; Atkinson et al., 1979: S-2-S-4].

It is worth noting that both Experimental and Control households did show an average improvement in numerous measures of neighborhood quality and concentration when they moved (even though there was no significant difference between the two groups). In contrast, Mayo et al. (1979) found, based on a sample of households in some traditional housing programs,

TABLE 10.3 Mean Change in Black Concentration for Experimental and Control Households

Treatment Type	*Black Households*	*White Households*	*Spanish-American Households*	*Total*
PITTSBURGH				
Experimental households	– 4.0	– 0.6	–	– 1.4
Standard deviation	23.2	7.5		13.0
(Sample size)	(211)	(698)		(909)
Control households	2.6	– 0.3	–	0.3
Standard deviation	16.8	7.1		9.8
(Sample size)	(63)	(254)		(317)
PHOENIX				
Experimental households	– 2.6	– 0.1	– 1.7	– 0.8
Standard deviation	23.8	7.5	14.1	11.7
(Sample size)	(52)	(438)	(207)	(697)
Control households	3.1	– 1.5	– 1.9	– 1.1
Standard deviation	26.3	8.5	7.7	11.3
(Sample size)	(27)	(180)	(69)	(276)

Note: Experimental/Control differences not significant at the 0.05 level in a two-tailed t-test.

SOURCE: Atkinson et al. (1979: Table 3-3).

TABLE 10.4 Mean Changes in Spanish-American Concentration for Experimental and Control Households

Treatment Type	*Spanish-American Households*	*White Households*	*Black Households*
Experimental households	– 4.0	– 0.8	– 2.1
Standard deviation	19.7	10.5	12.7
(Sample size)	(207)	(438)	(52)
Control households	– 4.8	– 1.6	– 0.9
Standard deviation	16.9	7.6	14.0
(Sample size)	(69)	(180)	(27)

Note: Experimental/Control differences not significant at the 0.05 level in a two-tailed t-test.

SOURCE: Atkinson et al. (1979: Table 4-3).

that moving into units in such programs meant an increase in economic and racial concentration, on average.

HOUSING CHANGE

Because the experiment covered two widely different treatment formulas, Percent of Rent and Housing Gap, the housing change of families receiving these two types of payment was analyzed separately (Friedman and Weinberg, 1978, 1979). In both cases, only movers showed increases in housing consumption (using several measures) significantly above that of otherwise similar Control movers.

The approach taken in analyzing the behavior of households in the Housing Gap treatment plans was to compare their housing change during the course of the experiment with that computed to be the normal change during that same time span (by comparison with Control households). Table 10.5 presents the estimated experimental effects on the two main measures of housing change used—housing expenditures and housing services (a measure of housing quality). It is clear that movers show significantly greater increases above normal in housing than do nonmovers. Indeed, only for one measure (services) in only one site (Phoenix) does any nonmover group show any increase significantly above normal.

The increases in rent above normal for movers are, for the most part, larger than those for all households. This result suggests that the overall response to a Housing Gap housing allowance would increase over time as more households move. The magnitude of the difference between movers and all households is not large, however, indicating that dramatic increases in average response subsequent to the first two years of any program are unlikely. There is evidence, however, that movers tend to dissipate more of their expenditure increase in price increases (as opposed to quantity or quality increases) than do nonmovers.[5]

Analysis of the housing response of Percent of Rent households took place within the context of housing demand analysis. The percentage of rent rebate can be analyzed as a direct

TABLE 10.5 Median Percentage Increase in Housing Expenditures and Housing Services Above Normal by Mobility Status Over Two Years

	Pittsburgh		*Phoenix*	
Household Group	*Percentage Change in Expenditures*	*Percentage Change in Services*	*Percentage Change in Expenditures*	*Percentage Change in Services*
ALL HOUSEHOLDS THAT MET REQUIREMENTS AT TWO YEARS				
Minimum standards households	4.3%	3.1%	16.2%**	10.2%**
	(2.7)	(2.5)	(3.9)	(3.7)
Minimum rent low households	2.8	0.0	15.7**	11.0**
	(2.5)	(2.0)	(4.4)	(3.8)
Minimum rent high households	8.5*	0.9	28.4**	18.0**
	(3.6)	(2.6)	(6.3)	(4.9)
Unconstrained households	2.6	3.4	16.0**	12.6**
	(3.1)	(2.5)	(5.6)	(4.7)
MOVERS THAT MET REQUIREMENTS AT TWO YEARS				
Minimum standards households	8.1	1.7	19.2**	7.6
	(5.3)	(5.1)	(5.5)	(5.3)
Minimum rent low households	5.1[a]	0.3[a]	14.5**	12.3*
	(4.6)	(4.3)	(5.5)	(5.4)
Minimum rent high households	14.0**	7.1	26.4**	13.9*
	(5.7)	(4.8)	(7.0)	(5.7)
Unconstrained households	3.7	9.2†	17.9*	13.4†
	(5.8)	(5.5)	(7.8)	(7.2)

TABLE 10.5 (Continued)

Household Group	Pittsburgh		Phoenix	
	Percentage Change in Expenditures	*Percentage Change in Services*	*Percentage Change in Expenditures*	*Percentage Change in Services*
STAYERS THAT MET REQUIREMENTS AT TWO YEARS				
Minimum standards households	1.4 (2.4)	1.0 (1.7)	3.1 (3.0)	5.4* (2.1)
Minimum rent low households	− 0.7 (2.3)	− 0.3 (1.5)	3.8 (4.3)	3.9 (2.4)
Minimum rent high households	2.1 (3.8)	− 3.4 (2.1)	[4.8] (10.2)	[2.6] (5.4)
Unconstrained households	0.5 (2.8)	0.1 (1.7)	[4.6] (4.4)	[6.7]* (2.5)

Notes: Sample sizes for housing services estimates are smaller than for expenditures due to extra data requirements. All numbers corrected for selection bias using Control households that did not meet the particular requirement at two years after enrollment except the expenditure increase for Minimum Standards households and all numbers for Unconstrained households. Brackets indicate amounts based on 15 or fewer observations. Standard error in parentheses.

[a]Correction based on 15 or fewer Control observations.

†t-statistic of estimated effect significant at the 0.10 level.

*t-statistic of estimated effect significant at the 0.05 level.

**t-statistic of estimated effect significant at the 0.01 level.

SOURCE: Friedman and Weinberg (1979: Table 7-33).

price reduction (see Friedman and Weinberg, 1978). As the price of housing is reduced more, a greater increase in housing consumption is expected. This pattern is confirmed for movers, but does not occur for nonmovers (Figure 10.4).

In order to estimate the extent of this responsiveness to price reductions in the cost of housing, demand functions were specified, having as independent variables, inter alia, the experimental rebate and average income. The entire sample of renters is not the best sample to use to estimate such demand functions. In theory, housing demand and expenditure functions are based on the household's choice of optimal, utility-maximizing amounts of housing under the implicit assumption that the search and moving costs of adjusting housing consumption to changed household circumstances are negligible. However, these costs may be significant. Households may not adjust immediately to correct imbalances in their consumption of housing and non-housing goods; thus, unless they have moved recently, they may not be consuming their desired amount of housing. Estimates of price elasticities based on all households may, therefore, not estimate the true elasticity accurately. This suggests that it is desirable to estimate a demand function solely for movers. Further, if renters generally adjust their housing by moving, then households that did not move would be expected to show little change in housing expenditures in response to the Percent of Rent rebates (see Figure 10.4). As these households move, they may well respond more like the households that moved during the experimental period. Thus, estimates for movers may provide a better estimate of the underlying demand function and the eventual response to a rent rebate than would estimates based on the entire sample.[6]

Table 10.6 presents the elasticities estimated for the sample of Percent of Rent and Control households moving between enrollment and two years after enrollment. The estimated point elasticities for price are -0.21 in Pittsburgh and -0.22 in Phoenix, while for income they are identical: 0.36 in both sites.[7] The site similarity is striking, suggesting that one demand equation could be estimated for the entire mover sample (pooled across the sites). When a site-specific intercept is allowed, the hypothesis that price and income elasticities are the same in the two sites is not rejected. These elasticities are in the low end of the range of estimates, but there appears to be little doubt that housing demand is price- and income-inelastic, with the absolute values of the elasticities likely to be less than 0.5.

FUTURE RESEARCH

The role of mobility is clear—it is the key to household response. Moving is the process by which households change their neighborhood, participate in the program, and increase

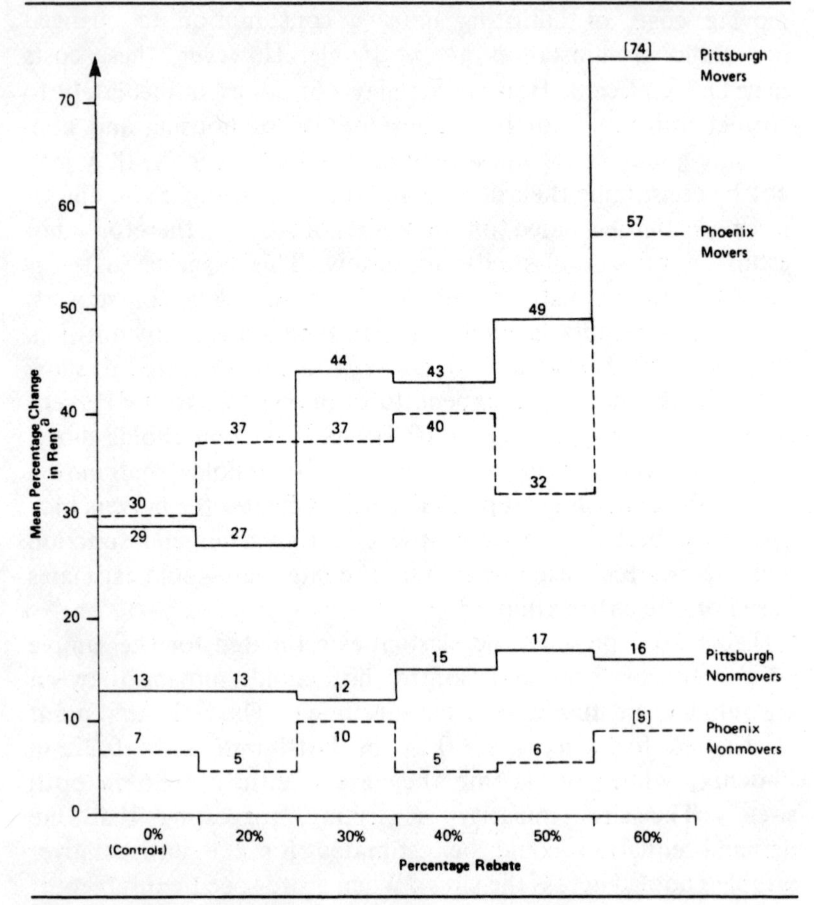

Figure 10.4 Mean Percentage Change in Housing Expenditures Between Enrollment and Two Years after Enrollment

SAMPLE: Percentage of Rent and Control households active at two years after enrollment, excluding those with enrollment incomes over the eligibility limits and those living in their own homes or in subsidized housing.

NOTE: Brackets indicate entries based on 15 or fewer observations.

[a]Percentage change in rent is defined as the mean of the ratio to the rent at enrollment.

SOURCES: Initial and monthly Household Report Forms.

their consumption of housing. Unfortunately, lacking in the analysis of the Housing Allowance Demand Experiment to date is a coherent model linking these outcomes with the experi-

TABLE 10.6 Price and Income Elasticity Estimates for Movers Sample

	Pittsburgh	*Phoenix*	*Pooled Sites*
Price Elasticity	−0.211**	−0.219**	−0.216**
	(0.063)	(0.059)	(0.043)
Income Elasticity	0.363**	0.364**	0.364**
	(0.052)	(0.042)	(0.033)
Sample Size	(236)	(292)	(528)

Note: Estimated using a log-linear form with three-year average income used as a measure of permanent income. Standard error in parentheses.

**t-statistic significant at the 0.01 level.

SOURCE: Friedman and Weinberg (1978: Tables 4-2 and 4-3).

mental incentives through the process of mobility. Of course, this linkage is implicitly present, as in Friedman and Weinberg's (1978) selection of movers to use for demand function estimates. Yet, the intimate linkages among the experimental treatments, participation in the experiment through finding adequate housing, mobility, and housing outcomes need to be made explicit.

Basic to such a model is a uniform theoretical framework for thinking about household behavior, a framework provided by microeconomics—in particular, that of the utility-maximizing household. In simple form, the household is assumed to maximize its utility U(H, Z), where H is housing and Z represents all other goods, subject to a budget constraint. The preexperimental household is viewed as consuming some possibly nonoptimal amount of housing and other goods, and must decide whether to change its current housing consumption in response either to a housing allowance payment—if it is eligible to participate at enrollment—or to the possibility of receiving such a payment—if it is not eligible.

A first step in formulating a unified framework was taken by Weinberg et al. (forthcoming). They developed a model of residential mobility and search that is tied directly to household utility through housing demand based in part on work by Quigley and Weinberg (1977). While not the comprehensive model of response called for above, as it deals only with the group of households already participating at enrollment, it is clearly on the right track. Their model is described briefly below.

Weinberg et al.'s model is cast explicitly in terms of a rigorous model of housing demand. This permits changes in household and housing market characteristics to influence a household's decision to search for and move to a different residence. The basic factor motivating the household is that of incentives to (benefits from) moving; the basic deterring factor is costs. In brief, moving is viewed as the process by which households adjust their housing consumption to their desired (equilibrium) levels. Households, therefore, search for housing and then move when the expected gains from changing their housing outweigh the costs of finding and moving to a new unit.

The existence of transaction costs of search and moving, including both monetary and psychological costs, immediately suggests that, having once chosen a satisfactory unit, households will not necessarily move when small changes render their current unit nonoptimal for their particular household characteristics. This immobility would lead to utility losses, in the sense that the household's level of utility attainable in the optimal unit would exceed the utility attained in the current unit (ignoring transaction costs). The household would move, however, only when the utility loss involved in staying in its current unit outweighs the costs associated with moving.

In principle, the utility gain foregone by a household in not moving to its equilibrium may be measured in monetary terms using the concept of the compensating income variation—the maximum amount of money households could spend on search and moving costs given the prevailing prices and income and still be as well off after the move as they were beforehand. If the compensating income variation is larger than the actual costs associated with moving, the household would, in theory, choose to move.

Using an estimated equilibrium demand function and the concept of consumers' surplus, Weinberg et al. quantify a measure of the disequilibrium between housing services actually consumed and the amount desired for Percent of Rent and Control households in the demand experiment. (Percent of Rent households automatically participate upon enrollment.) Three measures of household costs are also quantified: the out-of-pocket costs of moving possessions, the cost of searching for a new unit, and a proxy for the psychological costs of breaking neighborhood ties–namely, the rent discount associated with lengthy tenure in any one dwelling unit.

One particularly striking finding is the small magnitude of the potential benefits from moving which result from the price discounts offered by the demand experiment, especially in relation to the expected costs of moving. The average induced benefit averages from $2 to $3 per month. The small induced benefit from moving is a direct result of the low price elasticity of housing demand for low-income households. The economics of the housing demand of low-income households is such that relatively large changes in traditional "economic variables" such as prices and income result in relatively small changes in equilibrium housing demand. Further, because households that do not move to attain a new housing equilibrium are partially compensated for their suboptimal consumption of housing by the other goods and services that they are able to purchase by not moving, the magnitude of the benefit from moving is further decreased. In contrast, the costs of moving may be considerable. Many households pay below-market prices for their rental units as a result of either long-term tenancy, special deals between tenants and landlords, or simply mistakes in the prices charged tenants by landlords. The cost of surrendering such good deals may be considerable relative to the benefits; tenure discounts alone may be sizable enough to deter many moves.

Table 10.7 presents the results of the estimation of a benefit/cost model of mobility using logit analysis. Both the cost and the benefit measures perform well at both sites; the significant variables have the expected sign. In the combined site equation, almost all of the variables are both of the correct sign and highly significant.

TABLE 10.7 Disequilibrium Logit Model of Two-Year Mobility

Independent Variables	*Pittsburgh*	*Phoenix*	*Combined Sites*
Constant	0.5221 (0.4148)	0.9240* (0.3710)	1.1670* (0.1682)
Cost Measures			
Expected out-of-pocket moving costs	– 0.0150** (0.0057)	0.0147 (0.0183)	– 0.0178** (0.0027)
Expected search time	0.0047 (0.0049)	– 0.0128** (0.0048)	– 0.0042 (0.0032)
Current tenure discount	– 0.1119** (0.0148)	– 0.0838** (0.0119)	– 0.0965** (0.0090)
Benefit Measure			
Total disequilibrium	0.0047† (0.0027)	0.0043† (0.0025)	0.0047** (0.0018)
Chi-square of benefit measure[a] (significance)	2.51 (NS)	2.88 (0.10)	6.46 (0.05)
Proportion moving	0.342	0.548	0.431
Coefficient of determination (ρ^2)	0.07	0.09	0.10
Sample size	(672)	(513)	(1185)

Notes: Standard errors in parentheses below coefficient.
[a]With one degree of freedom.
†t-statistic significant at the 0.10 level
*t-statistic significant at the 0.05 level
**t-statistic significant at the 0.01 level
NS=not significant at the 0.10 level
SOURCE: Weinberg et al. (forthcoming: Table 3).

While these empirical results are encouraging, they are also somewhat disappointing. It is apparent that this model failed to capture the broad range of reasons for mobility. The explanatory power of the models is smaller than that of models that rely heavily on sociodemographic variables such as that of MacMillan (1978), although the number of explanatory variables is also smaller in Weinberg et al.'s model than in such alternatives. They propose several promising extensions of their work, including the use of demographically disaggregated demand functions to estimate benefit measures, decomposition of the housing bundle into components, and more rigorous specification of expected costs of moving.

The most important extension, however, is interactive determination not only of the demand functions and the mobility equations but also of residential location choice and program participation. Kennedy et al. (1977) have examined a model of participation cast in much the same terms as the model of Weinberg et al. They conceptualized the problem of participation as determining the minimum payment necessary to induce participation—that is, the payment which would make the household as well off participating as not. This, it is clear, is just a transformation of the concept of compensating income variation.

A comprehensive model must restate this concept in compensating income terms and formulate a joint simultaneous system of equations in which the following are represented:

(1) the probability of participation as a function of the compensating income variation of participating;
(2) the probability of moving as a function of the compensating income variation of moving; and
(3) the demand for housing, conceptualized as demands for separate housing components, one of which is location.

Development of such a comprehensive model requires a huge investment of time and resources in order to bear fruit. Such a model is conceptually akin to the models advocated by Lerman (1975), Weinberg (1977), and others, who suggest that numerous household decisions need to be modeled simultaneously, even if the state of the art is insufficiently advanced to estimate such a model. These decisions include residential mobility, choice of residential location, the quantity of housing services purchased, workplace mobility, and portfolio allocation (including home and auto purchase, the latter relating to the commuting decision).

Nevertheless, the results of the Housing Allowance Demand Experiment have been and will be used to further our understanding of the dynamics of housing choice and to help formulate housing policy, particularly with respect to the Section 8

Existing Housing program. The analysts of the demand experiment have provided a firm foundation upon which to continue such research. It remains to be seen whether a comprehensive model can be specified and subsequently verified by further empirical analysis of the Housing Allowance Demand Experiment data.

NOTES

1. This section is adapted from Wallace (1978).

2. A chi-square test showed it was possible to combine the two sites without significant loss of explanatory power if a site dummy variable was included. The higher probability rate in Phoenix appears to reflect regional differences in mobility (see MacMillan, forthcoming).

3. Shirley Mansfield has pointed out to me that this speculation usually did not take into consideration either the amount of the subsidy or the stringency of the housing requirements.

4. As pointed out by Peter Rossi, one problem with Atkinson et al.'s analysis was the clustering of households caused by the sampling procedure (elimination of high-income tracts). This may have precluded the observation of much "reverse mobility"–returning to areas of low-income or minority concentration. One other problem they faced was the possibility that the 1970 Census data they used to characterize neighborhoods was inapplicable to the experimental period (1973-1975).

5. See Friedman and Weinberg (1979: chap. 7) for more details.

6. See Friedman and Weinberg (1978), particularly Chapter 6, for an extended discussion of the issues surrounding this choice of movers. They present evidence that these issues do not in fact pose serious problems in estimating household responses to changes in price and income.

7. A linear expenditures equation was also estimated for each site. The mean estimated elasticities agree closely with the log-linear estimates. The functional form implies, however, that these elasticity estimates will increase (in absolute value) as income increases.

REFERENCES

ATKINSON, R., W. HAMILTON, and D. MYERS (1979) Economic and Racial Concentration in the Housing Allowance Demand Experiment. Cambridge, MA: Abt Associates.

FRIEDMAN, J. and D. H. WEINBERG (1979) Housing Consumption under a Constrained Income Transfer: Evidence from a Housing Gap Housing Allowance. Cambridge, MA: Abt Associates.

--- (1978) The Demand for Rental Housing: Evidence from a Percent of Rent Housing Allowance. Cambridge, MA: Abt Associates.

KENNEDY, S. D., T. K. KUMAR, and G. WEISBROD (1977) Participation under a Housing Gap Form of Housing Allowance. Cambridge, MA: Abt Associates.

LERMAN, S. R. (1975) "Location, housing, auto ownership and mode to work: a joint choice model." Prepared for the Fifty-fifth Annual Meeting of the Transportation Research Board.

MacMILLAN, J. (1978) Mobility in the Housing Allowance Demand Experiment. Cambridge, MA: Abt Associates.

--- (forthcoming) "The decision to move–evidence from the demand experiment." Occasional Papers in Housing and Community Affairs. Washington, DC: Department of Housing and Urban Development.

MAYO, S. K, S. MANSFIELD, W. D. WARNER, and R. ZWETCHKENBAUM (1979) A Comparison of Housing Allowances and Other Rental Assistance Programs Based on the Housing Allowance Demand Experiment, Part 1: Participation, Housing Consumption, Location, and Satisfaction. Cambridge, MA: Abt Associates.

QUIGLEY, J. M. and D. H. WEINBERG (1977) "Intra-urban residential mobility. a review and synthesis." International Regional Science Review 2: 41-66.

WALLACE, J. (1978) Preliminary Findings from the Housing Allowance Demand Experiment. Cambridge, MA: Abt Associates.

WEINBERG, D. H. (1977) "Toward a simultaneous model of intraurban mobility." Explorations in Economic Research 4: 579-592.

--- J. FRIEDMAN, and S. K. MAYO (forthcoming) "A disequilibrium model of housing research and residential mobility." Occasional Papers in Housing and Community Affairs. Washington, DC: Department of Housing and Urban Development.

11

Studying Residential Mobility: Administrative Records of the Housing Assistance Supply Experiment

MARK DAVID MENCHIK

□ THIS CHAPTER DESCRIBES a statistical procedure developed to analyze the mobility experience of households which participated in an experimental test of housing allowances. Although the procedure was designed for administrative records, it is generally applicable to other longitudinal microdata that are "censored" by short periods of observation. The procedure's application here shows the major role played by eligibility requirements in the mobility of housing allowance recipients. It also highlights a potential inferential problem that is widespread in the use of administrative records. Program requirements may cause self-selection of participants according to (in this case) predisposition to move. Self-selection may therefore bias comparisons with general populations.

THE SUPPLY EXPERIMENT'S HOUSING ALLOWANCE PROGRAM

The Housing Assistance Supply Experiment tests housing allowances for low-income households. The experiment is sponsored by the U.S. Department of Housing and Urban Devel-

AUTHOR'S NOTE: *This chapter derives from research supported by the U.S. Department of Housing and Urban Development under contract H-1787 and by the National Institute of Child Health and Human Development under grant RO1-HD12394. I am grateful for comments by Will Harriss and Peter Morrison. Conclusions are my responsibility.*

opment (HUD). This chapter examines the residential mobility of the 9000 households participating in the supply experiment which received allowances in the first two years of program operation. Unlike the Housing Allowance Demand Experiment, which samples allowance recipients, the supply experiment is open to all eligible persons in two midsize metropolitan areas: Brown County, Wisconsin, and South Bend, Indiana, whose central cities are Green Bay and South Bend.[1]

Most other housing assistance programs subsidize the occupants of specific dwellings either publicly or privately owned. Participants in HUD's experimental allowance program, however, choose their homes in the open market and, subject to the experiment's housing quality constraints, may move about and rent or buy homes as they prefer without affecting their allowance entitlement.[2] How many, then, will take their "portable" allowances and move?

A household may enroll in the allowance program if it satisfies requirements for income, assets, family composition, and residency. If, in addition, it occupies a certified dwelling, it will receive allowance payments. The income limit is set by the assistance formula, which calculates the allowance payment as the difference between the standard cost of adequate housing for a particular household size and a fourth of adjusted gross income. The asset ceiling was originally set high enough—$20,000 for nonelderly households, $32,500 for elderly ones—so that homeowners with low incomes would be eligible for assistance. Until August 1977, the family composition requirement excluded single-person households which were not elderly, disabled, or displaced by public action. The residency requirement is simply that the family be permanent residents of Brown or St. Joseph County.

To receive allowance payments, an enrollee must occupy a certified dwelling, either the one he resides in at enrollment (the enrollment dwelling) or one he moves into later. A certified dwelling must meet physical standards pertaining to the unit's state of repair and presence of certain minimum facilities (for example, a kitchen) and occupancy standards—no more than two persons per bedroom plus one room as a general living area for households of three or more persons. If the dwelling is rented, the enrollee and landlord must sign a year's lease. The housing allowance office staff tries to evaluate all enrollment dwellings, even if the enrollee is not interested in receiving allowance payments there (perhaps because he plans to move).

SCOPE OF STUDY

We investigated only moves out of enrollment dwellings (during the two-year period, recipients were enrolled an average of 11 months; in this short interval, less than five percent moved more than once). Further, only mobility within the metropolitan areas of the experimental sites is studied, since those who leave lose their eligibility for allowance payments and are thereafter not "at risk" of allowance-associated mobility.

This study examines only allowance program records. Designed primarily for administrative purposes, those records provide no direct information on residential mobility. Thus, the occurrence and timing of moves must be inferred from enrollment information, especially from evaluation records for each participant's potential and actual dwellings. Information on the 24 percent of enrollees who never received an allowance payment was inadequate even for inference. Enrollees who received payments have had at least one evaluation and therefore furnish the potential for inferred mobility information.

Limitation to administrative records has two consequences for this research. First, some of the recipients we studied never achieved certification of their enrollment dwelling; allowance receipt meant moving into another certified dwelling. For them, mobility was in part a program requirement, not a matter of personal choice. That fact influences our interpretation of the data. Second, allowance recipients were self-selected; that is, they chose to participate in the program and therefore do not constitute a random sample of eligibles. Self-selection may reflect a different propensity to move for recipients than for otherwise similar nonrecipients, which also argues for careful interpretation of the findings. This chapter's conclusions return to the matter of self-selection.

Enrollees whose enrollment dwellings are certified may stay in them or move out. If an enrollment dwelling fails the certification evaluation, the occupant has four options: (a) have it repaired successfully, (b) move to another dwelling that becomes certified, (c) move to an uncertifiable dwelling, or (d) stay in the uncertified enrollment dwelling. Only in the first two cases does the enrollee become a recipient and hence appear in our data. Of course, we have no mobility information for the last two.

If an enrollee whose enrollment dwelling was never certified later became a recipient, the enrollee must have moved. Such a move is more directly linked to program requirements than a move out of a certified dwelling. However, the allowance program does not force mobility from failed dwellings; the occupants can instead repair such dwellings or forego the allowance. Although administrative records exclude mobility intentions, some enrollees doubtless plan to leave their enrollment dwellings regardless of whether they are certified. Records show, in fact, that 86 percent of those who moved from uncertified to certified dwellings in the first two years made no attempt to repair their never-certified enrollment dwellings. (Of course, some dwellings are so deficient that neither the enrollee nor his landlord—if the unit is rented—could reasonably be expected to make successful repairs.) When eventual recipients leave their never-certified enrollment dwellings, questions such as the following arise: Are those moves a large fraction of program-associated moves, or, conversely, do most movers leave certified enrollment dwellings? How quickly do eventual recipients leave never-certified dwellings?

The short period during which the recipients are at risk of mobility limits the number of movers to be studied with our data. In the first two years of program operation, recipients were enrolled an average of less than a year, and only 21 percent moved. Distribution by site and factors such as residential tenure and certification status further divides the number of movers. For example, only 90 recipient homeowners (with complete records) moved in Brown County out of a total of 1602. About half had never-certified enrollment dwellings. Of the 1561 with certified enrollment dwellings, only 49 moved.

THE LENGTH-OF-STAY PROCEDURE

The data limitations here are typical of those faced by mobility analysts when working with administrative records covering short periods of time. Some few persons have moved, while others have not as yet relocated. How can analysis best capture mobility information from both movers and those who are as yet stayers? Traditionally, analysts have calculated simple mobility rates, dividing the number of moves by the aggregate period households were at risk of program-associated mobility. This is an oversimplification because households with long durations of residence (that is, periods at risk) display lower mobility rates (Speare et al., 1975).

Therefore, we developed a statistically efficient length-of-stay procedure that analyzes how long eventual movers stay in their dwellings. It uses information that simple mobility-rate analysis does not: the exact timing of the move, as well as its occurrence. Because of the length-of-stay equation resembles familiar regression equations, suspected influences on mobility can be tested as right-hand-side variables within the equation. That formulation therefore avoids decomposing the data into small subsamples.

The length-of-stay procedure can also analyze prospective mobility of households interviewed in surveys. Most surveys do not track households through successive residences. Consequently, the single or latest survey shows only that the household has not yet moved. The exact moveout date is censored information.

Even though movers' and stayers' information is different, our procedure uses both kinds of data to estimate a single equation. The characteristics of movers and other variables statistically explain the exact move-out date (their dependent variable). The period at risk of mobility is the dependent variable for stayers. The same explanatory variables explain why stayers' (prospective) move-out dates must follow our latest information about them. The supply experiment analysis shows, for example, that young recipients tend to be movers and older ones stayers. We do so by estimating the quantitative relationship between age of household head and length of stay—actual for movers and predicted for stayers.

The length-of-stay procedure has other advantages. It can summarize mobility either as lengths of stay or as probabilities of moving within stated periods of time. The latter are very much like traditional mobility rates, but with an important exception. Because our procedure models the duration-of-residence effect, it shows how mobility rates vary with the length of past residence.

DURATION-OF-RESIDENCE EFFECT AND THE FORM OF THE HAZARD FUNCTION

The length-of-stay procedure requires specifying how continued residence affects the chance of moving out. That specification is necessary in order to use both censored and uncensored data. Let ℓ indicate a household's length of stay, and $F(\ell)$ be the length-of-stay distribution function for a group of households. For any specified ℓ, the value of the distribution function is the probability that group members have stays shorter than ℓ (the distribution function also depends on the characteristics of the group members, but we suppress that detail for the present). The probability density function $f(\ell)$ is the derivative of $F(\ell)$. We may loosely think of $f(\ell)$ as the instantaneous probability of moving out exactly ℓ years after move-in.

The hazard function, $h(\ell)$, is the conditional probability density of moving out at time ℓ; that is, given at least that long a stay. Thus,

$$h(\ell) = \frac{f(\ell)}{1 - F(\ell)},$$

where $f(\ell)$ is the probability density of moving out at ℓ, computed on the basis of all households. $1 - F(\ell)$ is the fraction of households "surviving" to time ℓ—that is, the fraction of households moving out at ℓ or later.

The solid horizontal line in Figure 11.1 is the simplest possible hazard function, specifying that continued stay by itself neither increases nor decreases move-out probability. The decreasing hazard functions (dashed curves) represent a simple duration-of-residence effect: the longer one stays, the less likely one is to move. The single-peaked hazard function (curves combining long and short dashes) generalize the dashed curves. Note that the hazard function is a probability density, not a simple probability, and therefore may exceed one.

The highest probability of moving may not occur at move-in. Since housing needs and characteristics of residences change, even though a household is satisfied at move-in, the equilibrium between its preferences and its residences may gradually disappear. Dissatisfaction with the residence may mount because of the arrival of another child or an increase in income, for example. It may also take time after move-in to find another residence. In the allowance program, for example, some time passes between enrollment and evaluation and, perhaps, repairs. We can thus envision a variety of possible hazard functions consistent with particular theoretical views of what triggers the decision to move. We found that the postenrollment mobility of allowance recipients could best be phrased by "starting the clock" at enrollment: by studying a postenrollment stay in the enrollment dwelling.

Two practical considerations influenced the choice of a hazard function. With censored data, the hazard function cannot be estimated directly, so even a simple function may prove to be statistically intractable. Our hazard function made estimation feasible. Because different hazard functions require different estimating procedures, we chose a general function and thereby avoided developing different estimators. Limited experimentation with other shapes of the hazard function, moreover, showed that our hazard function best fit the data.

We assumed the hazard function was

$$h(\ell) = \frac{\alpha}{\ell(1 + \exp[-\alpha(\log\ell - \chi\beta)])} \qquad [1]$$

where χ is a vector of characteristics of the households and their enrollment residences, β is a parameter vector, and α determines the hazard function's shape. When α is one or less, the hazard function is everywhere decreasing. When α is greater than one, it

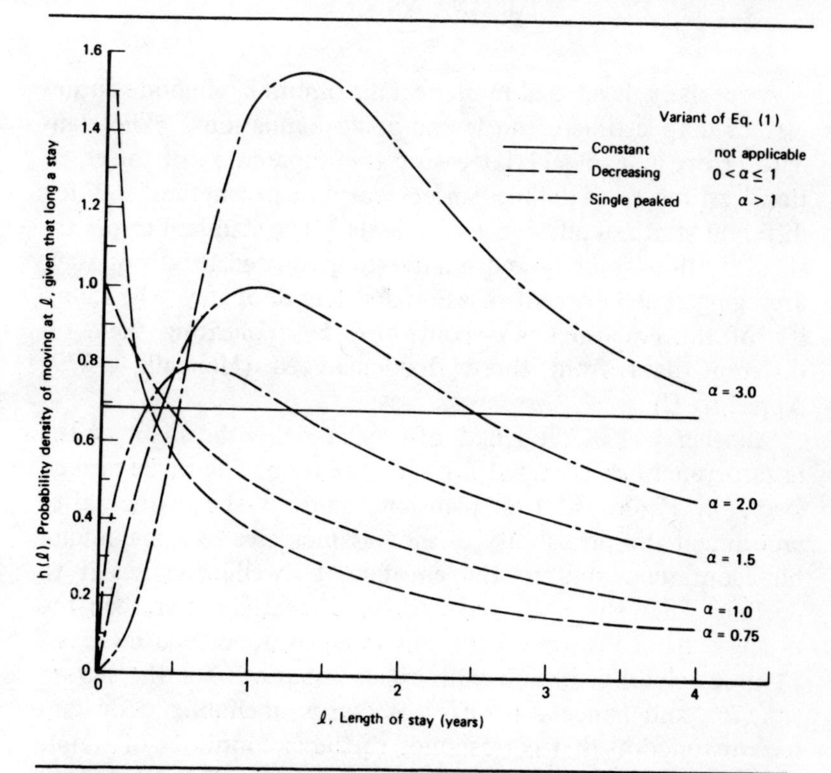

Figure 11.1 Theoretical Hazard Functions (Median Length of Stay, One Year)

has a peak. Figure 11.1 shows hazard functions which, although they all lead to a median length of stay of one year, have different values of α.

The assumed hazard function defines the following relationship between length of stay, characteristics of the recipient, and characteristics of his enrollment residence:

$$log\ \ell = \beta_o + \beta_1 x_1 + \beta_2 x_2 + \ldots + \beta_m x_m + \epsilon, \qquad [2]$$

where $x_1, x_2, \ldots, x_m$ are the recipient characteristics, and, as a consequence of our hazard function, the error term ϵ is a logistic random variable with mean zero and variance proportional to $1/\alpha^2$. This simple yet general equation parallels that of ordinary linear regression.

THE ANALYSIS

A specially developed numerical maximum likelihood routine was used to estimate the length-of-stay equations.[3] The standard errors in Table 11.1 measure the imprecision of the equations' estimates of the median postenrollment length of stay for different strata of allowance recipients.[4] The standard errors are small, both absolutely and relative to estimated length of stay, averaging at 9.8 percent of estimated length of stay. The stability of the equations was confirmed by replicating them on different data from those first analyzed (Menchik, 1979: Appendix C).

Another test of the length-of-stay model is the shape of the hazard function estimated for each equation. The single-peaked shape in Figure 11.1 is plausible, since it shows that after enrollment the probability of moving increases to a maximum, but continued stay in the enrollment dwelling causes it to decline. That shape fit the data significantly better than the other shape consistent with the model, monotonic decrease. Limited experimentation with other equations for the hazard function and hence still different shapes—including a constant hazard function that corresponds to the assumptions of simple mobility-rate analysis—also showed that the single-peaked shape fit the data best. Our estimation of the single-peaked hazard function and its empirical superiority to the constant hazard function is one sign that our procedure models mobility better than does simple mobility-rate analysis.

Table 11.2 lists move-out probabilities for periods of one, two, and five years after enrollment. An example will demonstrate how movement probabilities vary with postenrollment length of stay. Table 11.2 shows that the probability of a single-parent renter leaving his certified dwelling is .245 the first year after enrollment. Out of 100 such households, 75.5 are thus still staying in their enrollment dwellings at the beginning of their second year of enrollment. The two-year move-out probability is .500, so 50 of the 100 households remain at the end of two years. Consequently, 25.5 households are predicted to move within the second year (75.5 minus 50). That figure is .338 of the 75.5 households "surviving" to the beginning of the

TABLE 11.1 Estimated Postenrollment Length of Stay for Selected Groupings of Allowance Recipients, by Certification Status of Preenrollment Dwelling

	Length of Stay (years)					
	Enrollment Dwelling Certified		*Enrollment Dwelling Never Certified*		*All Dwellings*	
Type of Recipient	*Median*	*S.E.*	*Median*	*S.E.*	*Median*	*S.E.*
Renter	3.55	.38	.13	.01	2.72	.36
Elderly	7.39	.82	.14	.02	6.77	.82
Single parent	2.00	.09	.14	.01	1.41	.08
Other	2.39	.13	.12	.01	1.83	.12
Homeowner	18.79	1.85	.17	.03	18.05	1.85
All recipients	12.07	1.69	.14	.01	10.39	1.67

Note: S.E. is the approximate standard error of the median in the preceding column.

second year. The second year's mobility rate (.338) is considerably higher than that for the first year (.245). This shows the duration-of-residence effect as well as the danger of simple mobility-rate analysis assuming time-invariant rates.

The assumptions of this analysis should be kept firmly in mind, particularly when examining figures for homeowners whose dwellings were certified–both estimated lengths of stay and move-out probabilities, which were calculated from the same length-of-stay equation. Even though both replication and experimentation with different forms of the equation for that stratum showed the figures to be stable, they are based on data in which only two percent of the group was observed to move. Consequently, the estimates may be less reliable than for other groups.

Recipients whose enrollment dwellings were never certified moved very quickly, within a median of two months after enrollment. The equations for those strata contain generally different independent variables from the others.[5] The former include two aspects of program participation: whether or not the enrollment dwelling was repaired (repair lengthens stay) and the year of enrollment (those enrolled in the second year stay somewhat longer).

TABLE 11.2 Estimated Percentage of Recipients Leaving Enrollment Dwellings by Postenrollment Period

	Period after Enrollment					
	1 Year		*2 Years*		*5 Years*	
Type of Recipient	*Certified Dwellings*	*All Dwellings*	*Certified Dwellings*	*All Dwellings*	*Certified Dwellings*	*All Dwellings*
Renter	19.0	38.6	39.1	53.8	68.1	75.8
Elderly	8.2	16.0	17.3	24.4	38.3	43.6
Single parent	24.5	48.5	50.0	65.9	81.7	87.5
Other	21.1	40.4	43.4	57.3	75.4	81.4
Homeowner	2.3	6.2	5.6	9.4	16.3	19.6
All recipients	9.7	22.4	20.4	31.6	39.2	47.7

Figures 11.2 and 11.3 illustrate the differences in mobility among the various strata of allowance recipients. Eventual recipients leave never-certified dwellings so quickly that their hazard functions peak sharply (Figure 11.3). The probability of moving is understandably much less for those whose enrollment dwellings were certified (Figure 11.2). The more mobile strata in the second group have higher hazard functions that also peak earlier; for example, the functions for single-parent renters peak at 1.45 and 1.50 years after enrollment, while the flatter ones for elderly renters and homeowners peak at 1.62 and 4.88 years.

EFFECTS OF RECIPIENT AND RESIDENCE VARIABLES

Tenure and Life-Cycle Stage

After the certification status of an enrollment dwelling, tenure and occupant's life-cycle stage are the greatest influences on mobility. The latter influence, however, is limited to those whose enrollment dwelling was certified, because postenrollment length of stay does not vary appreciably among those with never-certified enrollment dwellings.

Renters whose enrollment dwelling was certified are eight times as likely as homeowners to move during the first preenrollment year; similarly, median stays for all renters are only 15 percent of those for all homeowners (Table 11.1). With an

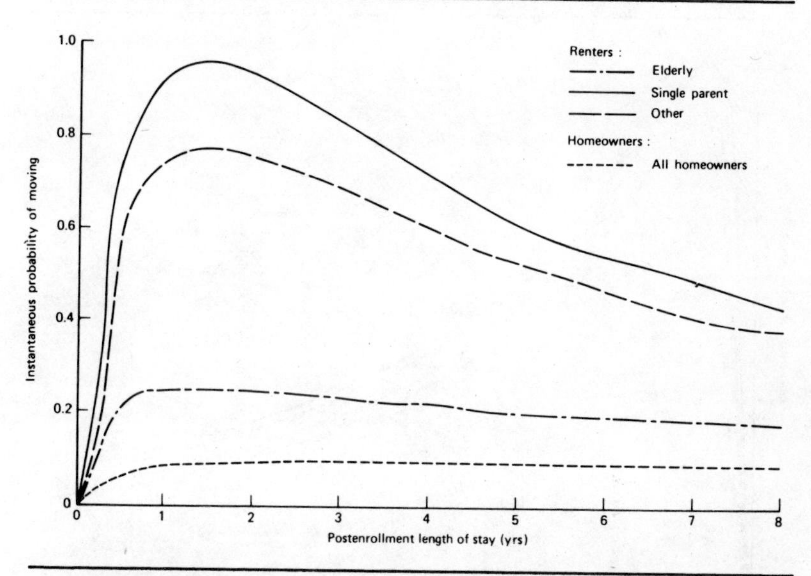

Figure 11.2 Estimated Hazard Functions for Recipients Whose Enrollment Dwellings were Certified

SOURCE: HAO records through June 1976 for Brown County and December 1976 for St. Joseph County.

estimated postenrollment stay of only two years, single-parent renters are the most mobile of all with certified enrollment dwellings. Elderly households are the least mobile of such renters, staying about 7.4 years after enrollment.

Our findings about life-cycle stage and tenure for recipients whose enrollment dwellings were certified parallel those for general populations (for example, see Quigley and Weinberg, 1977; Speare et al., 1975). The explanations advanced in that literature apply here, too.

Unlike renting, homeowning represents a large financial commitment, the more so because of the large transaction costs attendant on buying and selling. Homeowners are therefore more likely than renters to choose a residence that suits both their current needs and those of the foreseeable future. Homeowners who expect to stay in their new residence a long time may "put down roots" by modifying the house and grounds to their liking, establishing ties with their neighbors and local

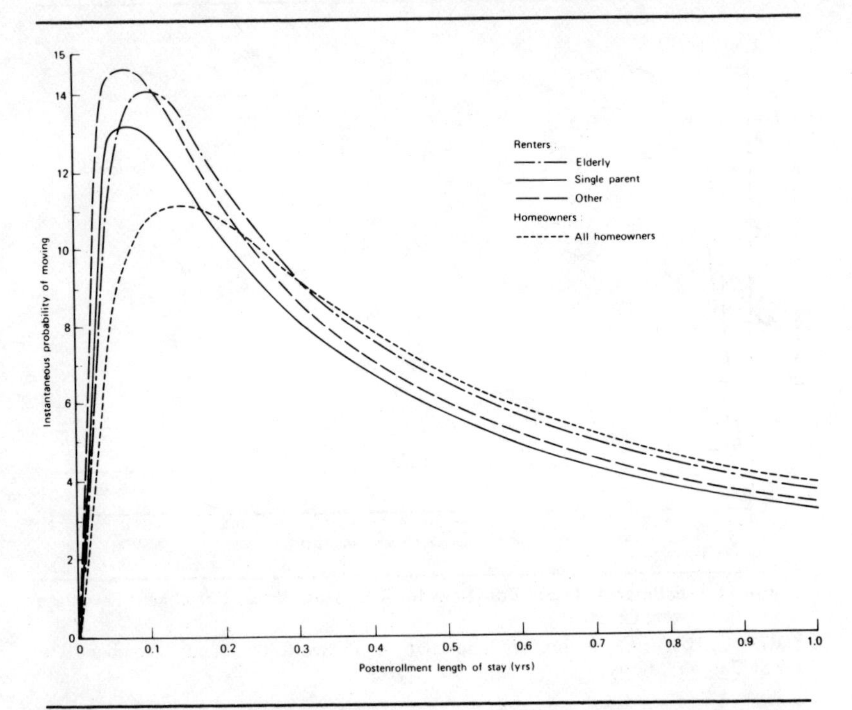

Figure 11.3 Estimated Hazard Functions for Recipients Whose Enrollment Dwellings were Never Certified

SOURCE: HAO records through June 1976 for Brown County and December 1976 for St. Joseph County.

organizations, and so on. Even when their residence no longer suits them, the difficulty and expense of selling the home may impede moving.

A household's life-cycle stage influences its mobility in several ways. In general, a household moves when its residence no longer adequately satisfies its preferences (including financial capabilities) and another one satisfies better. Households headed by young single or married persons change rapidly. Individuals marry; incomes alter; children are born and reach school age. A recently married couple may want a better home, particularly if the household now has two incomes. Similarly, new parents need more space for their children. As adults age, however, household change occurs less frequently.

Poor, single-parent families tend to live under frequently abruptly changing circumstances that compel frequent moves. Many such families result from the father's departure or death; their very formation therefore accompanies the loss of a wage earner, which drastically lowers their ability to pay for housing. If the mother's childcare responsibilities prevent her from working full-time, income is frequently an uncertain mixture of public subsidy, low-paying, part-time jobs, and perhaps child support payments or gifts from relatives. Consequently, it is not surprising that single-parent renter households are the most mobile of all allowance recipients.

Age of Household Head

Because homeowners were not disaggregated by life-cycle stage, their equations included instead the age of the household head in logarithmic form. Homeowners whose enrollment units were certified display an age elasticity of 0.94, with a standard error of .25.

Site and Race

In spite of the fact that vacancy rates were decidedly lower in Brown than in St. Joseph County, a recipient's site was not significantly associated with his mobility. The mobility influence of race disappeared when other characteristics were taken into account. The results of tests on other variables appear in Menchik (1979: 31-51).

EFFECTS OF PROGRAM-PARTICIPATION VARIABLES

We have seen that tenure and life-cycle stage exert strong, consistent influences on the postenrollment stay of recipients with certified enrollment dwellings, but that no demographic variable influences how long recipients stay in never-certified units. Program-related variables display an opposite pattern: Year of enrollment and whether repairs were done in the enrollment unit slightly affect the length of stay for those whose enrollment dwellings that were not certified, but have no effect on recipients with certified units.

Enrollment Dwelling Repair

Table 11.3 shows that most renters who repair a never-certified enrollment dwelling stay in it two or three times longer than otherwise, but still leave within a few months of enrollment. That effect was found for renters in all life-cycle stages, though not for homeowners. However, for certified dwellings, earlier length-of-stay equations showed that repair was a consistently nonsignificant influence on length of stay.

The variable role of repairs in mobility has no simple explanation. Perhaps those who unsuccessfully repair their failed units are more committed to their homes than others who do not attempt repair. The time taken for the repair itself may slightly lengthen their stay (although the time for repairs does not lengthen stays in eventually certified units).

Year of Enrollment

We sought evidence on whether recipients who enrolled toward the beginning of program operation had different mobility characteristics from those who enrolled later. No such differences were found for recipients whose enrollment dwellings were certified. A slight (but significant) difference appeared among certain renters who enrolled in the first or second years of program operation. Table 11.4 shows that single-parent and other never-certified renters who enrolled in the second year stayed about half again as long as those who enrolled in the first year. Year of enrollment had the same influence (or lack of influence) on length of stay when the sites were analyzed separately, and when time of enrollment was defined by half-year. Perhaps self-selection was operating. Those who enrolled in the first year may have needed (or wanted) allowances more strongly than those who enrolled later, and therefore acted quickly to satisfy program requirements.

Whether the enrollment dwelling was certified at first inspection did not explain subsequent mobility as well as did eventual certification status. Moreover, once information on the final outcome of certification and the presence of repairs was included in the analysis, the initial certification added no new

TABLE 11.3 Effect of Repair on Stay in Never-Certified Enrollment Dwellings

	Length of Stay (years)					
	Dwelling Repaired		*Dwelling Never Repaired*		*All Never-Certified Dwellings*	
Type of Recipient	*Median*	*S.E.*	*Median*	*S.E.*	*Median*	*S.E.*
Renter	.26	.06	.12	.01	.13	.01
Elderly	.34	.17	.12	.02	.14	.02
Single parent	.28	.05	.13	.01	.14	.01
Other	.21	.07	.12	.01	.12	.01
Homeowner	.14	.04	.20	.03	.17	.03
All recipients	.24	.06	.13	.01	.14	.01

TABLE 11.4 Postenrollment Stay by Year Enrolled: Recipients with Never-Certified Enrollment Dwellings

	Length of Stay (years) if Enrolled in:					
	First Year		*Second Year*		*Both Years*	
Type of Recipient	*Median*	*S.E.*	*Median*	*S.E.*	*Median*	*S.E.*
Renter	.10	.01	.15	.01	.13	.01
Elderly	*	–	*	–	.14	.02
Single parent	.10	.01	.16	.01	.14	.01
Other	.09	.01	.14	.01	.12	.01
Homeowner	*	–	*	–	.17	.03
All recipients	.11	.01	.16	.01	.14	.01

*Variable was nonsignificant; thus, length of stay was not calculated for each year.

information and therefore was nonsignificant. The same result occurred for a related variable: the number of housing deficiencies (either zero or a positive integer) found in the enrollment dwelling's first inspection. That variable is strongly associated with whether repairs were made and with the eventual outcome of certification; thus, it provided no new information.

It is now clear why we stratified recipients according to eventual rather than initial certification outcome. Not only is eventual outcome more strongly related to mobility, but stratification by initial outcome would combine two groups that, although each failed initial certification, had very different later mobility: those whose enrollment dwellings were never certified, and those who successfully repaired them. In fact, the latter group's mobility is indistinguishable from those whose enrollment dwellings were certified upon first inspection.

Notice, too, that because eventual recipients whose enrollment dwellings were never certified moved so quickly, their hazard functions differ greatly from those of other recipients. Estimation would therefore require stratification by final outcome even if the data were also stratified by initial outcome. Explaining mobility by both initial and final outcome would therefore add no more to our results than would stratification by final outcome alone.

RELATION BETWEEN POSTENROLLMENT AND PREENROLLMENT STAY

One reason that different life-cycle and tenure groups have different postenrollment lengths of stay is their different preenrollment stays. On the one hand, recipients who occupy their enrollment dwellings a long time before enrolling tend to stay a long time afterward; for example, elderly renters occupy certified enrollment units a median of four years before and seven years after enrolling. On the other hand, single-parent renters stay less than a year before enrolling and only two years thereafter.

Postenrollment length of stay thus increases with preenrollment stay for certified dwellings. The relation holds both across and within the four recipient strata, although the relation is, of course, weaker within groups. Our findings agree with the duration-of-residence effect observed in general populations. Three explanations may account for these facts in the recipient population. First, those who stay in their home a long time before enrollment may do so because they are intrinsically less mobile than others; if so, they are also less likely to leave after enrollment. Second, a long stay may strengthen ties to home and neighborhood. Third, since long-standing renters enjoy a rent discount, which they lose on leaving, they may resist moving.

CONCLUSIONS

LENGTH-OF-STAY PROCEDURE

The new procedure efficiently analyzes heterogeneous mobility histories censored by a short period of observation. Allowance recipients' periods of enrollment (and thus period at risk of program-associated mobility) varied greatly, from mere days to two years, but averaged less than a year. During this period, only eight percent of recipients moved whose enrollment dwellings were certified for allowance receipt. Half of all recipients were homeowners with certified enrollment dwellings; of these, less than two percent moved. On the other hand, virtually all eventual recipients whose enrollment dwellings were never certified moved within a few months. Consequently, the data were disaggregated into eight strata according to the prime mobility determinants (other mobility influences appeared as righthand-side variables). Even after disaggregation, mobility was accurately and reliably estimated. Standard errors averaged less than ten percent of estimated length of stay, and the length-of-stay equations were successfully replicated on an independent data set.

Many of the estimated hazard functions are importantly nonhorizontal, showing mobility as strongly influenced by duration of residence after enrollment. This duration-of-residence effect causes analysis based on simple mobility rates to be inaccurate. In all cases, hazard functions belonged to the same family of single-peaked curves where the propensity to move first rises with continued residence and later declines.

INFLUENCES ON RECIPIENTS' MOBILITY

The primary determinant of an allowance recipient's mobility is whether his enrollment unit was ever certified. In the first two years of the program in both sites, only 14 percent of all recipients' enrollment dwellings failed certification. However, since recipients with never-certified enrollment dwellings must move to a certified dwelling to receive an allowance, that group accounts for 68 percent of all the recipients who left their enrollment dwellings. Recipients with certified dwellings are estimated to stay in them a median of 12 years after enrollment; those with uncertified enrollment dwellings stay in them only 0.14 years after enrollment (that is, less than two months).

For recipients whose enrollment dwellings are certified, mobility is influenced largely by personal characteristics and tenure. Homeowners stay a median of 19 years; renters, less than four years. Life-cycle stage also influences mobility. Single-parent families who are renters are the most mobile, staying about two years in their enrollment units, whereas elderly renters stay about seven years after enrollment.

The few influences acting on recipients' brief stay in never-certified enrollment units suggests that program characteristics play the important role, whereas personal or residential characteristics influence both the mobility of the general population and the mobility of recipients with certified enrollment units. Tenure and life-cycle stage do not influence the point at which recipients leave never-certified enrollment dwellings. On the other hand, stay is roughly doubled for renters who unsuccessfully attempt repairs, although it is still less than four months. Repairs to an eventually certified unit did not affect the resident's mobility. Some renters stayed slightly longer in never-certified units if they enrolled in the second year of program operation. No such association holds for those in certified enrollment units.

SELF-SELECTION

If recipients self-select according to predisposition to move, the measured mobility of recipients will be influenced by that predisposition, not solely by program requirements or observable characteristics of the recipients or their housing. Two examples illustrate how self-selection might operate.

First, some eligible persons' homes may not be certifiable even with a reasonable amount of repair. Knowing that they must move to receive allowance payments, those who are strongly attached to their homes may never apply. If so, self-selection would screen from the population studied those who are disinclined to move. By implication, eventual recipients whose enrollment dwellings are not certifiable are those who remain—that is, persons relatively *less* reluctant to move. They may enroll fully expecting to move—which could explain the short stays of recipients in never-certified enrollment dwellings. (On the other hand, those households may simply want an allowance strongly enough to move quickly if their enrollment dwelling is not certified.)

Second, another self-selection process may operate. If a household lives in a residence of standard quality and wants to stay, it may apply for an allowance in hopes of getting help with housing expenses. Although that possibly may explain the much longer stays of recipients in certified enrollment units, allowance payments may actually have no effect on their mobility decision; or allowances may simply enable them to stay longer than otherwise.

Assessing true causal influences of enrollment unit certification on mobility requires mobility data about enrollees who never receive payments, as well as about eligible persons who never apply. Mobility histories before enrollment would also be helpful, since preenrollment stays in the enrollment dwelling may signal attachment to that unit. Even better would be data on the length of stay in the dwelling occupied before the enrollment unit, which can indicate how mobile the household really is. Such data could be supplied by the supply experiment's surveys, which interview a sample of households, some of whom participate in the allowance program.

Fortunately, the data analyzed here allow us to assess the first example of the self-selection hypothesis: that those whose enrollment dwellings were never certified did not stay in them long before enrolling, were unattached to them, and chose not to have the units certified. Table 11.5 shows that (controlling for tenure and life-cycle stage) those whose enrollment dwellings were never certified usually stayed in them about as long

TABLE 11.5 Preenrollment Stay in Certified and Never-Certified Enrollment Dwellings

	Median Preenrollment Stay (years)		
Type of Recipient	*Enrollment Dwelling Certified*	*Enrollment Dwelling Never Certified*	*All Dwellings*
Renter	1.04	.77	.96
Elderly	3.72	4.38	3.83
Single parent	.65	.67	.66
Other	.80	.60	.76
Homewner	10.97	8.06	10.89
All recipients	4.00	.99	3.18

before enrollment as those whose enrollment dwellings were certified.[6] By that measure at least, failure to certify, not prior tendency to move, causes eventual recipients to leave never-certified dwellings so quickly.

NOTES

1. Details of the two experimental sites and of the data appear in Menchik (1979). which describes the analysis in more detail.

2. Unlike the demand experiment, homeowners may participate in the supply experiment.

3. The estimating routine can be found in Menchik (1979: 28-30).

4. The following conventions apply to the length-of-stay estimates: (a) The household is assumed to continue to be at risk of program-associated mobility, that is, to continue to be enrolled; (b) mobility determinants other than those appearing explicitly in a table are set equal to the mean for that particular group of allowance recipients.

5. For estimation, the data were divided into eight strata corresponding to the curves in Figures 11.2 and 11.3.

6. There is a slight tendency for never-certified "other" renters and homeowners to stay a shorter time before enrollment than their counterparts in certified dwellings. However, that disparity cannot explain the enormous difference in postenrollment length of stay according to certification status of the enrollment dwelling.

REFERENCES

MENCHIK, M. (1979) Residential Mobility of Housing Allowance Recipients. N-1144-HUD. Santa Monica, CA: Rand Corporation.

QUIGLEY, J. M. and D. H. WEINBERG (1977) "Intra-urban residential mobility: a review and synthesis." International Regional Science Review 2: 41-66.

SPEARE, A., S. GOLDSTEIN, and W. H. FREY (1975) Residential Mobility, Migration and Metropolitan Change. Cambridge, MA: Ballinger.

Part IV

The Political Context

☐ THE CHAPTERS in the previous two sections represent the orthodox academic approach to residential mobility in the sense that the analyses take the institutional context as given and build their arguments around the paradigm of individual choice under exogenously defined constraints. Process is seen in terms of the exercise of individual preferences over a set of available alternatives subject to the limitations of budget and information. Such an approach applied to public policy issues can lead to the accusation that it is "only able to give technical recommendations for the realization of pre-given ends" (Duncan, 1976: 11). It provides no sense of the larger social context within which the particular functioning of the urban system is to be explained; nor, in more pragmatic terms, does it lead to an appreciation of the kinds of politically realizable constraints which are likely to be effective.

Such a critique leads to a different perspective on how to establish links between residential mobility and public policy. It suggests that we consider, in a more fundamental way, the processes by which different groups within society gain, or are granted access to, housing and related services. Only against such a background can we understand the consequences of past public actions and the potential efficacy of those proposed.

The chapters by Michael Dear and Norman and Susan Fainsteins develop this position in detail. Dear's paper begins with the observation that the urban areas of many cities are experiencing increasing concentrations of service-dependent populations, creating what he calls "The Public City." In seeking an explanation of this phenomenon he presents a strong case for moving outside the usual choice paradigm. It is not that discharged mental patients or users of services for the handicapped do not make locational choices, but that the constraints, both of an inclusionary and exclusionary nature, are sufficiently complex to diminish markedly the utility of such a conceptualization. Furthermore, as Duncan's comments imply, reduction of this group segregation is not simply a matter of changing administrative rules. As has been discovered by those involved in attempts to alter zoning ordinances to permit groups homes in residential neighborhoods, preserving the home environment dominates the desire to help disadvantaged groups, at least in the political arena. Even if policy objectives can be agreed upon, the problems of designing and implementing effective programs depend on a deeper understanding of social and political realities than is implied by most conventional analyses of mobility.

In discussion, Susan Fainstein made the critical point that American society has relied on "opportunity and motion, privatism and competition, not community and stability" to sustain its way of life. Two consequences follow from this observation. First, there is no rigid social law which specifies how social groups are spatially distributed within cities; rather, the way in which different groups gain access to housing means that social desirability, at least in part, is defined by where the rich and the poor live. If we use public programs to manipulate the quality of dwellings and associated amenities in specific neighborhoods, a likely outcome is that the more affluent will move into the area and thereby make it more desirable. In other words, the underlying social dynamic produces a predictable redistribution as a consequence of public action. A second point is that the current emphasis on neighborhood and community stability embedded in recent federal legislation stands in contradiction to the societal role of mobility; if the general premise is maintained that stability is a desirable goal, then the use of such place-orientated programs as a mechanism for improving the housing conditions of low-income households must be subject to serious questioning.

Although the above arguments are often developed within the framework of a critique of capitalist society, it is important that the issues are not dismissed because the ideological stance is seen as antagonistic. Whatever one's ideological position, the following chapters still make a strong case for evaluating the role of mobility within a broader institutional context which imparts social and political meaning to movement behavior. Such contexts and associated meaning vary both historically and spatially; for example, in contrasting the strength of neighborhood-based movements in Europe and the United States, it is important to assess whether observed differences in these movements reflect different expectations of and experiences with stability and mobility as well as variations in the development of grassroots politics. Each society at particular times has its own internal dynamic which needs to be understood as defining the context for design and implementation of public policy.

REFERENCE

DUNCAN, S. S. (1976) "Research directions in social geography in housing opportunities and constraints." Transactions of the Institute of British Geographers 1 (New Series): 10-19.

12

The Public City

MICHAEL DEAR

☐ THIS IS AN EXPLORATORY ESSAY on the origins and consequences of spatial concentrations of service-dependent populations and helping agencies in the inner city. This tendency toward a "public city" has only recently become a focus of attention; its meaning–indeed, its very existence–has been the subject of dispute. However, there now seems to be sufficient evidence to warrant a comprehensive analysis of the public city. In this chapter, special attention is paid to its role in meeting the residential location needs of the service-dependent populations. The essay proceeds from a simple empirical basis which attempts to establish the nature and dynamics of the public city. From this, the housing needs of the service-dependent are defined; in the third section, the underlying process of residential differentiation is explained as the functional basis for the public city. Finally, the major policy consequences of the public city are discussed.

DIMENSIONS OF THE PUBLIC CITY

MENTAL PATIENTS AND THE PUBLIC CITY: HAMILTON, ONTARIO

The current trend toward community-based mental health care began in the mid-1960s. Impetus for the movement derived from several sources, most notably the beliefs that long-term incarceration in hospitals did more harm than good and thereby infringed upon civil liberties; that community-based care would aid in the resocialization of the disturbed patient; and that cost savings could be achieved through transfer of patients to treatment settings outside the hospital. In short, an alliance between civil rights libertarians and fiscal conservatives was achieved. Attainment of their objectives was facilitated by contemporary advances in psychiatric treatment, particularly in chemotherapy, by which the extremes of behavior disorders could be controlled. As a consequence, a massive shift in treatment setting has occurred: mental hospitals have been closed, or reduced in scale, and many community mental health facilities opened. The last decade has witnessed a strong increase in active caseloads of community-based facilities and, to a lesser extent, of psychiatric units in general hospitals. In contrast, the active caseload of mental hospitals has declined to about one-third of its 1965 level. At the same time, a much larger turnover in the psychiatric hospital population is reflected in the growing volume of admissions and discharges (Dear, 1977).

The move toward community-based care is having a clear impact on the social structure of downtown Hamilton, an industrial city of over 300,000 population at the western edge of Lake Ontario. Its core area is typical in its relative abundance of unemployed, poor, elderly, and service-dependent populations (Social Planning and Research Council, 1977). This core is an area of 29 census tracts with 98,849 people in 1976, a decline of 16 percent since 1971.

Between April 1, 1978 and March 31, 1979, the Hamilton Psychiatric Hospital admitted 413 people who gave residential addresses in the city of Hamilton. Of this total, 61 percent came from the core area (Figure 12.1). During the same period, 495 people were discharged from the hospital, 70 percent to destinations in the core area (Figure 12.2). The geographical concentration of the discharged population was confined predominantly to a nine-tract area close to the central business district (CBD), and included a significant proportion (12 percent) of people

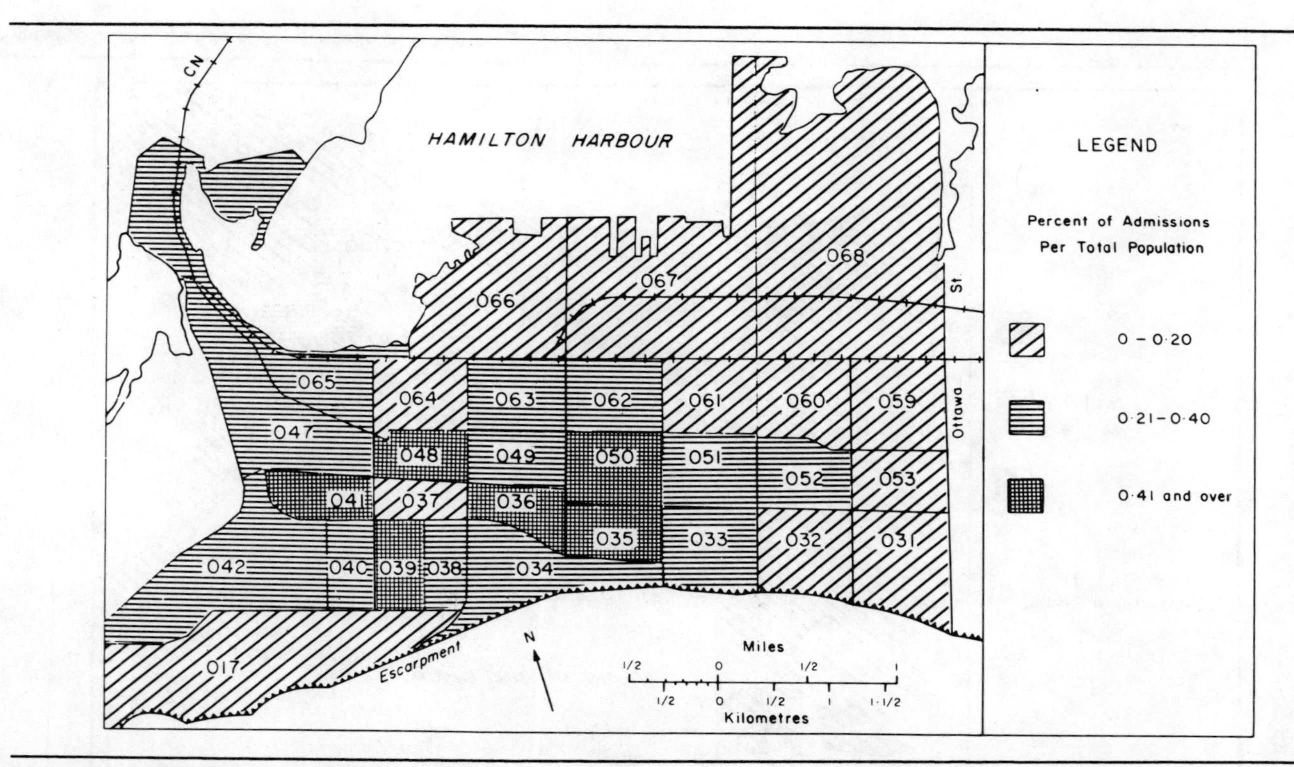

Figure 12.1 Distribution of Admissions to H.P.H.

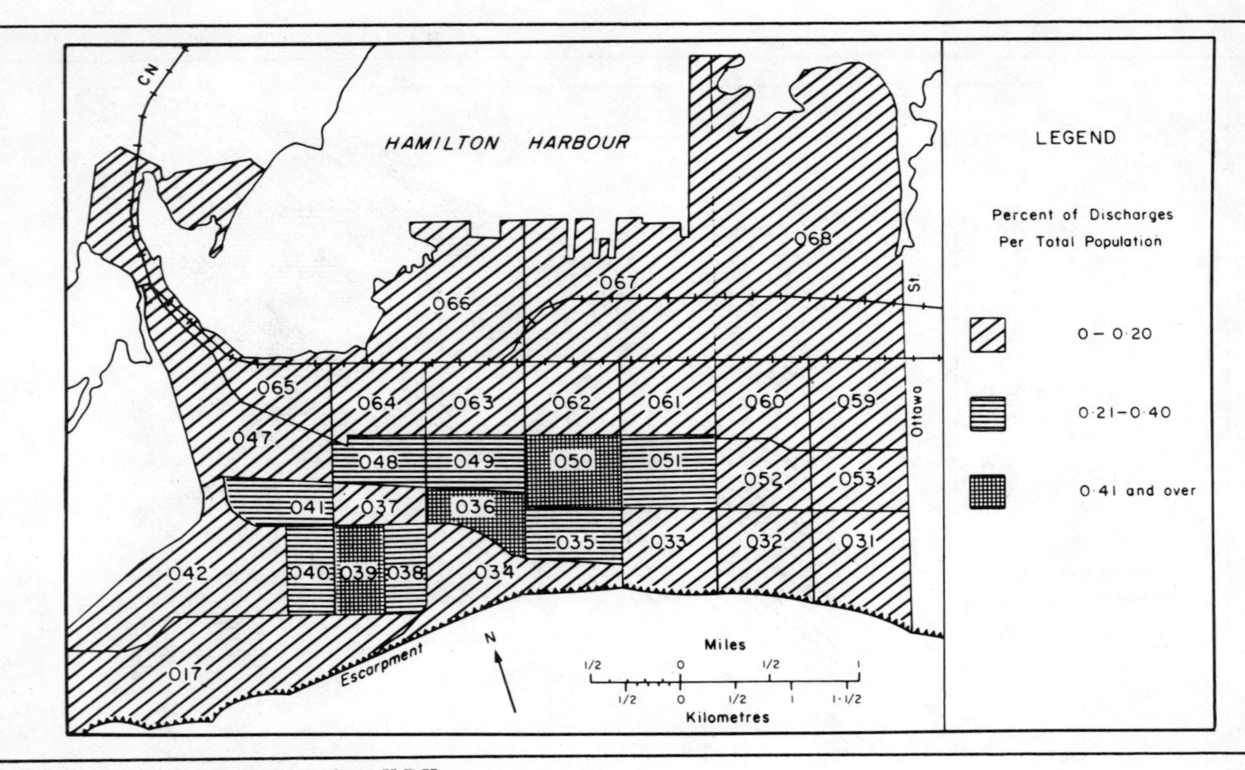

Figure 12.2 Distribution of Discharge from H.P.H.

who originated from outside Hamilton on admission. It is also significant, from the viewpoint of the helping professions, that the majority of those discharged to the core suffered from the more chronic psychiatric disorders. For instance, 27 percent of the core discharge group was schizophrenic, compared with only 7 percent of the rest-of-the-city discharge group.

One of the major attractions of the inner city for this group is the concentration of aftercare services in the core (Figure 12.3). This is especially true for those patients requiring continued supervision in boarding homes or continued medication in outpatient clinics. However, on many occasions when a patient returns to his or her family, some community support system is still required. Such facilities, for the most part, are located in core areas. Three main reasons have been given for this (Dear, 1977):

(1) the formal assignment of mental patients to institutional aftercare facilities which tend to proliferate in downtown locations because of planners' actions and because of the availability of large convertible properties;
(2) the extent of community opposition to mental health facilities in other neighborhoods; and
(3) an informal process of spatial filtering by which a volatile majority of mobile patients tends to gravitate toward the transient areas of rental accommodation in the inner urban core.

Only the last of these points requires further elaboration here. For present purposes, it is important to note that the discharged patients are extremely mobile. In a previous follow-up study of a cohort of 169 Hamilton patients, no less than 25 percent had made at least one move within six months of discharge. The movers were traced through a sequence of addresses: nine merely changed location; seven were readmitted to the hospital; and 21 were "lost." The losses are typical of the problem patient, who probably filters down into the transient ghetto situation. They did not leave forwarding addresses, left without notifying the landlord, and were often alcoholic. Of the 21 patients, four were lost after release from the criminal justice

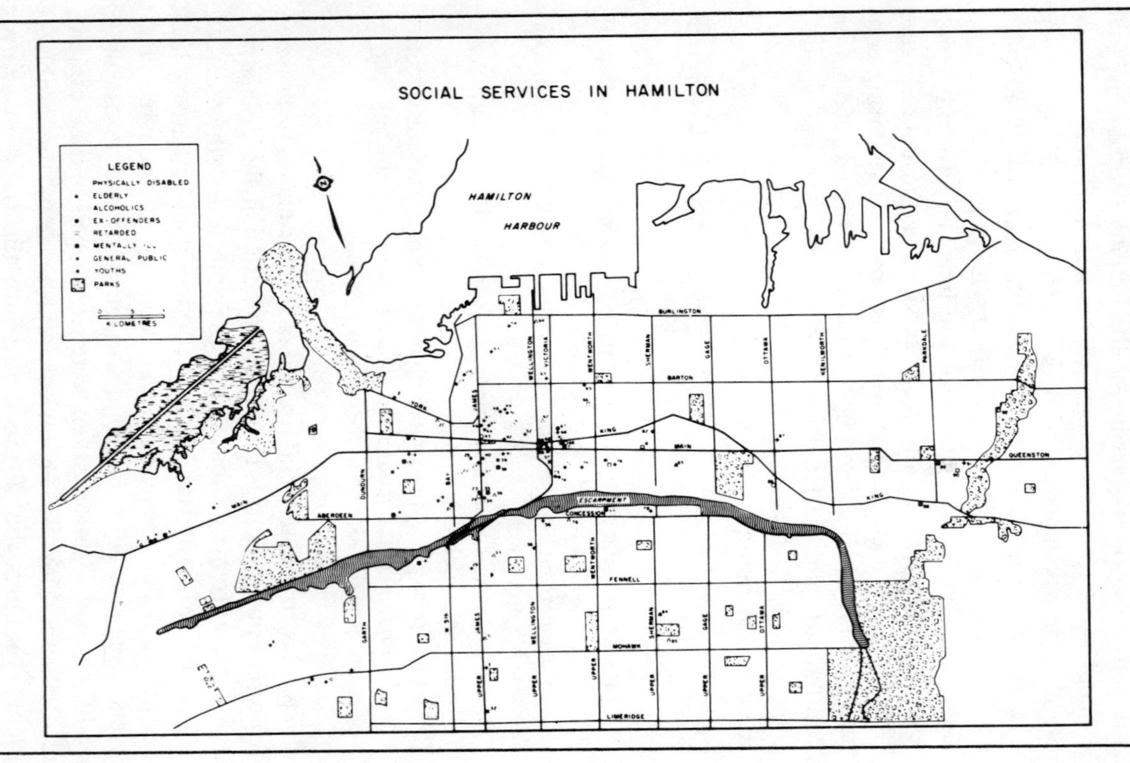
SOCIAL SERVICES IN HAMILTON
LEGEND
PHYSICALLY DISABLED
ELDERLY
ALCOHOLICS
EX-OFFENDERS
RETARDED
GENERAL PUBLIC
YOUTHS
PARKS
KILOMETRES
HAMILTON
HARBOUR
BURLINGTON
BARTON
KING
MAIN
YORK
QUEENSTON
ESCARPMENT
CONCESSION
ABERDEEN
FENNELL
MOHAWK
LIMERIDGE
DUNDURN
BAY
JAMES
WELLINGTON
VICTORIA
WENTWORTH
SHERMAN
GAGE
OTTAWA
KENILWORTH
PARKDALE
GARTH
UPPER

Figure 12.3

system; 10 dropped out of sight immediately; and seven were lost only after an extensive search through subsequent addresses. It is worth noting that only three out of the 21 were discharged to their families; the majority of losses were discharged in their own recognizance (that is, to self-care: Dear, 1977).

In summary, a new "asylum without walls" seems to be developing in downtown Hamilton. Since the operation of the formal institution has been reduced, its population has transferred itself to the inner city. The core area thus becomes a reservoir for hospital admissions and the major discharge destination for an increasing proportion of people treated in hospitals. The growth of service agencies to aid the discharged patients intensifies the attractive qualities of the core, and a self-reinforcing cycle is thus initiated, intensifying the growth of the "public city."

OTHER EMPIRICAL EVIDENCE

There is very little empirical evidence on the nature and structure of the public city, although there are extensive literatures on public facility location (see the bibliography by Freestone, 1977) and on the locational behavior of the urban poor (Gorham and Glazer, 1976). Only recently, however, have aspects of service-dependent populations and helping agencies been considered simultaneously. For example, White (1979) emphasized the locational interdependencies which exist among various service types. In his analysis of the "co-locational" patterns of mental health, mental retardation, and other social support services, he observed two discrete (but overlapping) tendencies. One strong agglomeration was defined for mental health and linked services, and a second, looser agglomeration for social support and public health services. White emphasized that traditional planning designs for a decentralized ("accessible") facility set ignore the potential benefits of an agglomerated set. This theme is echoed in several studies of discharged mental hospital patients. These studies suggest that the service facility agglomeration may provide a highly supportive environment for the disadvantaged. Within the "ghetto," ex-patients help each other to find accommodation, locate job opportunities, or run local newsletters (Wolpert and Wolpert, 1976).

For another type of service-dependent population, the nonworking poor, Wolch (1978) found that the pattern of human service facilities in Philadelphia affects, and is affected by, the location of poor nonworking households. Both households and facilities are concentrated in the oldest, most deteriorated inner-city neighborhoods. Elsewhere, Wolch (1979) suggested that the journey-to-service replaces the journey-to-work as a major expense of the household budget. Hence, the residential location decision of service-dependent households is determined by the distribution of location-specific service "hubs." A recursive cycle of colocation is thus initiated, whereby services and the service-dependent coalesce to form service-dependent "ghettos" in the inner city.

Some doubt has been cast by Coughlin et al. (1976) on the special attention being paid to the spatial concentration of facilities. They examined the spatial distribution of three agency categories in Philadelphia: administrative, direct-service (common to many groups), and direct-service (targeted to specific groups). They found that the service facilities were far from evenly distributed throughout the city. The observed facility concentrations were associated with the needs of the residential population, but specific relationships with community instability were not identified. Smith (1975) also questioned the utility of the patient-agency "ghetto" for the patient.

However, it seems incontrovertible that there is a trend toward the concentration in the inner city of a host of service-dependent populations and professional agencies intended to help them. These two elements are acting in concert to create a new social structure, the public city. What is uncertain are the dimensions of this public city. Over half of the mental hospital population in the United States now receives care "in the community"; Bassuk and Gerson (1978) estimated that the hospital population has declined from 550,000 in 1955 to under 200,000 at the time of their study. We know that the mentally ill have been joined in the public city by many other service-dependent groups, including the elderly (Golant, 1975; Wiseman, 1978). It was recently suggested that one-quarter of the population in declining U.S. central cities was dependent upon public sector programs of one kind or another (Wolch, 1979). Moreover, Wolpert (1978) estimated that 61 service facilities in New Brunswick, New Jersey, attracted over 6000 people daily to the city's core, exceeding the number of daily shoppers and private-sector work trips to the CBD.

HOUSING NEEDS OF THE SERVICE-DEPENDENT

The "service-dependent" population encompasses an enormous variety of needs, reflecting the degree of client disability. For example, one person may be totally disabled physically and permanently confined to a wheelchair. Another individual may suffer a stroke and require intensive hospital care for a short period before returning to everyday living. The different needs of various types of service-dependent clients require a complex, heterogeneous set of service opportunities. It should also be noted that any client may need to move rapidly through a range of service opportunities as his or her treatment proceeds. Both these characteristics have a profound effect on the housing needs of the special population being discussed in this essay. These effects are best seen by constructing a taxonomy of service clients and alternative treatment settings or residential opportunities. In the following section one such taxonomy, devised by Dear and Wolch (1979), is reviewed.

A TAXONOMY OF SERVICE CLIENTS AND TREATMENT SETTINGS

A basic taxonomy of human service clients encompasses a continuum which extends from total client dependence to complete client autonomy. At the former end of the spectrum, the client is totally dependent on the human service system for life support; at the other extreme, the client is totally self-supporting and requires no assistance from the service sector. The objective of the service system is to move clients toward the autonomy end of the continuum—that is, to encourage client participation in the full experience of life, including schooling, recreation, family, and community. On the other hand, the autonomy-dependence (A/D) concept recognizes that certain groups will possess long-enduring handicaps which may vary in their intensity (thus providing for movement in both directions along the continuum) and which may require life-long dependence on the service system.

In principle, all members of a given client group could be allocated along the A/D continuum. However, a comprehensive taxonomy requires the use of multiple continua in order to include the mental, physical, and social dimensions of service need. In all three cases, complete dependence implies that a client is totally reliant upon the human service system for life-support; such would be the case with those who are hospitalized as chronically mentally ill, the profoundly mentally retarded, or the severely physically disabled. At the other end of the continua, complete autonomy implies exit from the service system.

Treatment settings for human service clients may be characterized along a single dimension representing a continuous progression from closed, protected environments to open, unrestricted environments. In general, position along this dimension indicates the degree of control exerted by the provider of services over client behavior and activity. More specifically, for any given client subgroup, position on the protected-unprotected (P/U) dimension implies three things: a specific facility type, a particular setting, and an associated treatment regimen. Thus, many maximum-security institutions for criminals or the chronically mentally ill envisage a completely closed, protected environment for total life-support of the inmates. Treatment occurs in a relatively large-scale facility divorced from any community base. At the other end of the spectrum are services which act as a buffer between the outside world and the service-dependent population approaching independent living. Such informal programs usually imply no specific service site, but some services such as home visit programs may still be provided to assist clients in reentry into the community. These services are clearly predicated on close links with the social networks of existing communities.

Between the two extremes lie a variety of community-based services which are related to established social networks but are clearly part of the formal service network. These are usually provided from small-scale, designated facilities and may include welfare agencies and specialized retirement communities at the relatively unrestricted end of the taxonomy; or they may include group homes and homes for special care at the more protected end. It is important to note that client "normalization" implies progress along the P/U dimension over time. Hence, a criminal may proceed from total institutional care, through an open prison, to a community-based readjustment and parole program. Conversely, recidivism can lead to regression toward the protected end of the taxonomy.

The goal of any human service agency is to match a client with an appropriate treatment setting. In operational terms, this implies that the degree of client "envelopment" be minimized, such that the client undergoes the least possible removal in time and space from normal activities, concomitant with treatment requirements. Allocation to an overly protective environment could cause frustration and withdrawal; allocation to a too-open environment can place too great a strain on a client and may result in renewed breakdown.

RESIDENTIAL LOCATION CHOICE OF SERVICE-DEPENDENT POPULATIONS

Given the needs of the service-dependent populations, it is clear that any approach to their residential location behavior must be highly unconventional. Wolpert (1978) suggested that the journey-to-service facility may replace the journey-to-work in traditional models, and Wolch (1979) rewrote conventional utility functions to include receipt of "in-kind" income and the costs of community opposition. This latter essay indicated that residential choice is severely constrained by client service needs, service opportunities, and the exclusionary practices of state and community. It would seem, therefore, that the conventional approach to choice behavior in residential location is not representative of the majority of service-dependent populations. There are many reasons for this assertion: (1) Service-dependence is generally involuntary and unpredictable; the decision to become dependent is usually taken on one's behalf by some other person. (2) For clients to seek care themselves requires recognition of symptoms and an acceptance of the "sick" role. (3) Except in a minority of cases, clients are rarely required to make direct payment for service; hence, the concept of "purchasing" services is often redundant. (4) Consumer behavior in the marketplace is essentially irrational, due to the inability of clients to assess the quality of service received and their inability to choose among service providers, largely through lack of appropriate knowledge. (5) Finally, human services have a wide range of external benefits which accrue to society as a whole; for example, it is clearly in society's interest that a potentially harmful patient be detained in custodial care. One important consequence of such external effects is that non-consumers have an important voice in the kinds of services provided.

In an "economy" dominated by welfare recipients, it is not surprising that the response on the supply side has been ambiguous. The difficulties involved in measuring service outcomes and, thus, of developing appropriate pricing policies have effectively removed profit as a factor in many service sectors. Even when private care is possible, strict regulation and licensing procedures are generally involved. Without effective demand and a traditional supply-side response, the "market" for human services remains undeveloped. It has been replaced by a public decision mechanism. For present purposes, the most important consequence of collective action is that decisions on service delivery tend to be based upon the concept of client need rather than demand. Need is a complex notion, embracing political and professional judgments about the preferred level of well-being for a client population. It is not something that can be determined in the open market.

In summary, the service-dependent population tends to be directed toward service and/or residential opportunities by some "other" decision maker. Only in exceptional cases does the residential location behavior of the service-dependent begin to approximate that of the normal population. Hence, the origin of the "ghetto" of service-dependent in the public city must lie somewhere in people's status as service-dependent.

RESIDENTIAL DIFFERENTIATION AND THE CREATION OF THE PUBLIC CITY

In this section, I examine the assertion that the public city is the outcome of urban collective action; that it is not some arbitrary creation resulting from the aggregation of many individual service-dependent decisions, but a structural feature which is both functional and convenient in contemporary urbanization. In order to show this, three different elements must be examined: first, the significance of the historical coincidence of abandonment and deinstitutionalization; second, the functional utility of residential differentiation in capitalist cities; and third, the alliance between community and state in creating the public city.

ABANDONMENT AND DEINSTITUTIONALIZATION

The notion that our inner cities act as a "host" for the service-dependent, the poor, and the deviant populations is not new. The earliest systematic attempts at social area analysis associated with the Chicago sociologists in the 1930s clearly identified the functional importance of the inner core for disadvantaged groups. Several decades of factorial ecologies have confirmed the persistence of this structural feature. However, many observers have recently commented upon fundamental changes in the pattern of urban structure in North American cities. These trends have been summarized by Adams (1976), among others. One major theme which has been identified is the process of commercial and residential abandonment of inner-city neighborhoods. The continued suburbanization of residential and employment opportunities without sustained immigration has, as its necessary corollary, an extensive abandonment of the obsolescent structures of the inner core (Grigsby, 1963). However, this normal market process has been exacerbated by a "cycle of abandonment" which is associated with a complete collapse of confidence in the regenerative capacity of the inner-city housing market. A contagious process of abandonment has taken hold in many older cities, representing a deep-seated, structural response to the pattern of contemporary suburbanization (Dear, 1976). The fiscal crises of American cities have only aggravated these trends (Alcaly and Mermelstein, 1977) and ensured that inner-city redlining becomes a normal finance and realtor response to depressed market conditions.

It is at this point that two important historical trends intersect to create a totally new dynamic for structural change. First, abandonment in our older cities is leaving a vacuum which was formerly taken up by continued inmigration and population growth. Now, however, the massive population migration to the older industrial cities is almost absent, and no land user of equivalent dimensions is moving in to take its place (Sternlieb and Burchell, 1973). In many cities, the tendency for governmental or quasi-governmental institutions to occupy this space has been the only factor preserving the physical and social fabric of the abandoned core areas (Dear, 1976). Second, a major change has occurred in the care of service-dependent populations during the past two decades. As we have seen, there has been a strong move away from institution-based care and toward community-based care. This shift has affected most service-dependent groups, including the mentally disabled, elderly, juvenile delinquents, and ex-prisoners. These groups have joined the poor, unemployed, welfare recipients, and single-parent families in the competition for space in the inner city. In short, a vacuum has been created in the inner city, and the movement toward deinstitutionalization has created a population ready to fill it.

RESIDENTIAL DIFFERENTIATION

Neither abandonment nor deinstitutionalization, taken separately or together, provide the necessary and sufficient conditions for the growth of the public city. Since urbanization is a concrete social event, embedded within a specific social order, the growth of the public city ultimately must be sought in some wider theory of society. Let me begin this search with a simple axiom: *The growth of differentiated neighborhoods in North American cities must (somehow) be functional for the capitalist economy, and must therefore reflect the social organization of capitalism.*

In capitalism, individual status and power is defined with reference to control, or lack of it, over the means of production. The existence of differential individual market capacities based on ownership of property, on educational or technical skills, or on manual labor-power is the source of class structuring. Two factors are important in the structuring of class relations: the mediate and the proximate (Giddens, 1973: chap. 6). The *mediate* factors are governed by market capacities and the distribution of mobility chances in society, since the greater the limits on mobility, the more likely are identifiable classes to form. The lack of intergenerational movement reproduces common life experiences, and such homogenizing of experience is reinforced by limitations on an individual's mobility within the labor market.

The effect of "closure" generated by the mediate structures is accentuated by the *proximate* factors of class structuring, according to which the basic within- and between-class structures are intensified and further differentiated. These more "derivative" characteristics are generated by the need to preserve the process of capital accumulation (Harvey, 1975). There are three groups of proximate factors.

(1) *The division of labor* within capitalism which is a force for both consolidation and fragmentation of class relationships. It favors the formation of classes according to the extent to which it creates homogeneous groupings. On the other hand,

the profit-motivated drive for modernization and efficiency often implies a specialization of labor functions and, hence, a fragmentation within an otherwise homogeneous group.

(2) *Authority relations* are a second force for class structuring. These may occur as a hierarchy of command with the productive enterprise, although, as Harvey (1975: 359) emphasized, it is equally important that the nonmarket elements in society be so ordered that they sustain the system of production, circulation, and distribution.

(3) The third source of the proximate structuring of classes, *distributive grouping,* is an aspect of consumption rather than production. Distributive groupings are those relationships and their concomitant status implications which involve common patterns of consumption of economic goods. They act to reinforce the separations initiated by differential market capacity, but "the most significant distributive groupings . . . are those formed through the tendency towards community or neighbourhood segregation" (Giddens, 1973: 109). This tendency is based on many factors, including income and access to the mortgage market, and ultimately gives rise to distinct "working-class" or "middle-class" neighborhoods.

Transformations in the social structure thus inevitably imply transformations in spatial structure. For instance, the simple precepts of class structuralism tend to lead to the creation of relatively homogeneous environments in terms of social status. Such symbolic differentiation of urban space reflects choice in associates and opportunities for interaction in a class-differentiated society. The fundamental theoretical issue is the extent to which spatial form determines, or is determined by, social forces. Castells (1976: 77) argued that technical progress reduces the role of space as a determinant: It is not that space is external to the social structure and unaffected by it, but that its specific importance may be diminishing. However, Soja (1978: 10) suggested that once the organization of space is regarded as a purposeful social product, then it can no longer be regarded as a separate structure with its own rules of construction and transformation independent of social practice. The production of space is both an ideological and a political process which is a vital mediator in capitalist production and reproduction (Lefebvre, 1970).

In his analysis of inequality and poverty, Peet (1975) developed the formal links between social and spatial theory. Having shown that inequality and poverty are endemic to capitalism, Peet proceeded to demonstrate how space encourages their reproduction through generations. Central to his thesis is Hagestrand's notion of a "daily-life environment," composed of residence and/or workplace and defined by the physical friction of distance and the social distance of class. Each social group operates within a typical daily "prism," which, for the disadvantaged, closes into a "prison" of space and resources (Peet, 1975: 568).

Deficiencies in the environment, particularly limitations on mobility and the low density and quality of social resources, must clearly limit an individual's potential, or market capacity; similarly, low income limits access to more favorable environments. A self-reinforcing process thus sets in, and it is easy to understand how an individual can carry an "imprint" of a given environment and how the daily-life environment can act to "transmit" inequality.

In reproducing the ensemble of sociospatial inequality and poverty, capitalism naturally produces a class-differentiated society, each stratum of which is allowed to reproduce itself using varying proportions of its income to raise the next generation (Engels, 1972). Since the amount of money spent by each stratum varies, unequal resource environments are produced which help to perpetuate the class system. The city is thus composed of a "differentiated hierarchy of resource environments which reflect the different hierarchical labour demands of the capitalist economy" (Peet, 1975: 569). A similar approach to spatial residential differentiation was taken by Harvey (1975). He emphasized that differentiated space is an outcome of the capitalist production and reproduction processes, and not, as more typically construed, a product of the aggregate of individual consumer preferences.

Community and State

The processes by which residential differentiation creates the public city are indirect. They are not directly induced by the market, since the market for service-dependent populations is almost totally dysfunctional. Instead, the public city is formed through residential forces of community exclusion and through state intervention in the form of urban planning policy.

Community attitudes and behavior toward the service-dependent appears to depend upon resolution of an interdependent perceptual "trade-off." On one hand, there is a positive psychological benefit of helping a group in need; on the other, there is a negative, protectionist attitude toward one's daily-life environment. This confusion of motives was evident in a recent study of community attitudes toward mental illness in Toronto. Community attitudes seemed to resolve themselves into four factors: (1) authoritarianism, which implied a view of the mentally ill as an inferior class requiring coercive handling; (2) benevolence, a paternalistic, kindly view of patients derived from humanistic and religious principles; (3) social restrictiveness, viewing the mentally ill as a threat to society; and (4) a community mental health ideology, reflecting an anti-institutional bias in care philosophies (Taylor et al., 1979).

The service-dependent, like other minority groups such as the poor, are restricted in their selection of residence, workplace, and recreational outlets. Their continued isolation can be interpreted as part of a wider system of sociospatial organization which causes the separation of antagonistic groups. Thus, just as the processes of residential differentiation cause the appearance of class-separated and ethnically separated neighborhoods, so similar processes tend to isolate and exclude the service-dependent. The power of sociospatial exclusion operates at two separate levels in the community: in relation to the individual and the group. First, the dependent person is subject to a series of informal and formal exclusionary forces which operate at the individual level. Informally, a disability often tends to make the individual distinguishable in a social setting. Moreover, individuals have been observed to make personal behavioral adjustments to exclude the offending individual. More formally, organizational exclusion can occur, as when an individual is disciplined for aberrant behavior in the workplace (Scheff, 1967). Second, and more important for our present purposes, is the set of mechanisms of group exclusion. This refers to the generic ability of communities to exclude undesirable or noxious objects and people from their neighbourhoods. In an early study of exclusion of the mentally ill, for instance, Aviram and Segal (1973) recognized several strategies used by communities to place "social distance" between them and the mentally ill. These included formal strategies such as the use of legal ordinances (especially zoning), and informal strategies, such as physical abuse of facility or client.

If it is true that a limited environment of social resources has a significant impact on one's life chances, then it is evident that the household has an enormous stake in the local environment. Hence, the need becomes paramount to protect one's environment from any undesirable negative impact. It seems likely that the entrance of the service-dependent into a community is perceived as a threat to the environmental resource base of the neighborhood and hence the market capacities contained within it. Accordingly, the community's power for spatial exclusion is often marshalled to prevent their incursion.

The state is also implicated in the exclusion of the service-dependent. De facto planning policies have been devised in response to increasing community opposition. Locational strategies are chosen to minimize conflict over siting decisions. Hence, while neighborhoods with political clout may exclude "deviant" groups, other, less powerful neighborhoods are saturated with service facilities (Wolpert et al., 1975). For those populations which depend upon incomes derived directly or indirectly from political decisions, the influence of state intervention is all-pervasive. Locational choice tends to be preempted by one's service-dependent status. Moreover, the state is responding to those groups which have most to contribute toward easing the crises of capitalism; those whose problems have few ramifications for the social order tend to receive little, if any, attention (Roweis, 1980). This is in keeping with the state's long-term policy objective of crisis management (Habermas, 1976; Offe, 1976).

DISCUSSION

There can be no conventional conclusions at the end of such a speculative essay as this. I have attempted to focus on the origins of the spatial concentration of service-dependent populations and their associated helping agencies. That such a public city is developing no longer seems in dispute. What is of more interest is why it has arisen and how it relates to the developing urban social structure. This essay has suggested that a wide variety of service-dependent needs exist and that they typically will not be met without public intervention in the market.

The public city is part of the process of residential differentiation which is a functional component of capitalist urbanization. Two important historical trends intersect to facilitate this growth of the public city: the residential and commercial abandonment of obsolescent inner cities, and the rapid deinstitutionalization of service-dependent populations. Community practices of exclusion and state planning policies are implicated in the growth of the public city.

What, then, are the important policy consequences of the public city? Two issues seem to be of overriding concern. The first concerns the costs and benefits of the public city. The concentration of clients and facilities seems to provide a supportive environment for the user, but this environment is dominated by society's wounded. From the exclusionary community's viewpoint, however, the concentration of "deviants" in the transient, variegated city core probably seems the least threatening solution. The state, for its part, continues its pragmatic approach to caring for its dependents while responding to the claims of the powerful in urban society. The helping professions, hardly mentioned in this chapter, are fighting to control the flood of deinstitutionalization and seem barely able to rise above the daily responsibilities of service provision. Hence, first and most simply, we must address the question of whether this flood of service-dependent people into the city is a good or a bad thing.

Second, related to the organic growth of the public city is the question of the optimal spatial pattern for human service facilities. Recent attempts to "undo" the ghetto of ex-mental patients in California, New York, and Ontario seem designed to prevent saturation and to share the burden of care among all neighborhoods. This is being done without clear planning guidelines on the relative merits of a centralized or decentralized service system. The provision of a comprehensive aftercare support system for service-dependent populations, offering a choice of residential opportunities equivalent to that of the "normal" population, would require a much greater level of state involvement than would appear tolerable, especially in times of public expenditure restraint. Thus, the second question: What is the optimal pattern of community-based care, and to what extent can it be provided by the state?

REFERENCES

AVIRAM, V. and S. P. SEGAL (1973) "Exclusion of the mentally ill." Archives of General Psychiatry 29: 126-131.

ADAMS, J. S. [ed.] (1976) Urban Policymaking and Metropolitan Dynamics. Cambridge, MA: Ballinger.

ALCALY, R. E. and D. MERMELSTEIN [eds.] (1977) The fiscal Crisis of American Cities. New York: Vintage Books.

BASSUK, E. L. and S. GERSON (1978) "Deinstitutionalization and mental health services." Scientific American 238: 46-53.

CASTELLS, M. (1976) "Theory and ideology in urban sociology," pp. 60-84 in C. G. Pickvance (ed.) Urban Sociology: Critical Essays. London: Tavistock.

COUGHLIN, R. E., K. BIERI, and T. PLANT (1976) "The distribution of social services within the city of Philadelphia." Discussion Paper Series No. 93. Philadelphia: Regional Science Research Institute.

DEAR, M. (1977) "Psychiatric patients and the inner city." Annals, Association of American Geographers 67: 588-594.

--- (1976) "Abandoned housing," pp. 59-99 in J. S. Adams (ed.) Urban Policymaking and Metropolitan Dynamics. Cambridge, MA: Ballinger.

——— and J. WOLCH (1979) "The optimal assignment of human service clients to treatment settings," pp. 197-210 in S. M. Golant (ed.) The Location and Environment of Elderly Population. Washington, DC: V. H. Winston.

ENGELS, F. (1972) The Origin of the Family, Private Property and the State. New York: International Publishers.

FREESTONE, R. (1977) Public Facility Location. Council of Planning Librarians Exchange Bibliography 1211.

GIDDENS, A. (1973) The Class Structure of the Advanced Societies. London: Hutchinson University Library.

GOLANT, S. M. (1975) "Residential concentrations of the future elderly." Gerontologist 15: 16-23.

GORHAM, W. and N. GLAZER (1976) The Urban Predicament. Washington, DC: The Urban Institute.

GRIGSBY, W. (1963) Housing Markets and Public Policy. Philadelphia: University of Pennsylvania Press.

HABERMAS, J. (1976) "Problems of legitimation in late capitalism," pp. 363-387 in P. Connerton (ed.) Critical Sociology. New York: Penguin.

HARVEY, D. (1975) "Class structure in a capitalist society and the theory of residential differentiation," pp. 354-369 in R. Peel, M. F. Chisholm, and P. Haggett (eds.) Processes in Physical and Human Geography. London: Heinemann.

LEFEBVRE, H. (1976) "Reflections on the politics of space." Antipode 8: 30-37.

OFFE, C. (1976) "Political authority and class structures," pp. 388-421 in P. Connerton (ed.) Critical Sociology. New York: Penguin.

PEET, R. (1975) "Inequality and poverty: A Marxist-geographic inquiry." Annals, Association of American Geographers 65: 564-571.

ROWEIS, S. (1980) "Urban planning in early and late capitalist society," in M. Dear and A. J. Scott (eds.) Urbanization and Urban Planning in Capitalist Society. New York: Methuen.

SCHEFF, T. J. [ed.] (1967) Mental Illness and Social Process. New York: Harper & Row.

SEGAL, S. P. and U. AVIRAM (1978) The Mentally Ill in Community-Based Sheltered Care. New York: John Wiley.

SMITH, C. J. (1975) "Being mentally ill–In the asylum or the ghetto." Antipode 7: 53-59.

Social Planning and Research Council of Hamilton (1977) "A Socio-Economic Atlas of the City of Hamilton." Hamilton, Ontario, Canada.

SOJA, E. W. (1978) "Topian marxism and spatial praxis: a reconsideration of the political economy of space." Presented at the annual meeting of the Association of American Geographers, New Orleans.

STERNLIEB, G. and R. W. BURCHELL (1973) Residential abandonment: The Tenement Landlord Revisited. New Brunswick, NJ: Center for Urban Policy Research, Rutgers University.

TAYLOR, S. M., M. DEAR, and G. B. HALL (1979) "Attitudes toward the mentally ill and mental health facilities." Social Science & Medicine 130: 281-290.

WHITE, A. N. (1979) "Accessibility and public facility location." Economic Geography 55: 18-35.

WISEMAN, R. F. (1978) "Spatial aspects of aging." Resource Papers for College Geography 78-4. Washington, DC: Association of American Geographers.

WOLCH, J. (1979) "Residential location and the provision of human services." Professional Geographer 31: 271-276.

--- (1978) "Residential location of service-dependent households." Ph.D. dissertation, Princeton University. (unpublished)

WOLPERT, J. (1978) "Social planning and the mentally and physically handicapped," pp. 95-111 in R. W. Burchell and G. Sternlieb (eds.) Planning Theory in the 1980s. New Brunswick, NJ: Center for Urban Policy Research, Rutgers University.

--- and E. WOLPERT (1976) "The relocation of released mental hospital patients into residential communities." Policy Sciences 7: 31-51.

WOLPERT, J., M. DEAR, and R. CRAWFORD (1975) "Satellite mental health facilities." Annals, Association of American Geographers 65: 24-35.

13

Mobility, Community, and Participation: The American Way Out

NORMAN I. FAINSTEIN and SUSAN S. FAINSTEIN

□ AMERICANS have traditionally emphasized individual mobility rather than class action as the appropriate response to inequality. This formulation has important consequences for geographical mobility, political community, and public policy. While federal programs now claim to emphasize neighborhood improvement for the benefit of present residents, such an outcome seems to depend on maintaining the extant neighborhood social structure. In other words, it requires the creation and/or maintenance of community in at least its most limited sense of a stable residential grouping. However, American values and investment dynamics stymie the development of such a community.

In the United States social and geographical mobility tend to converge, thereby reproducing social inequality in physical terms. For the affluent, success is measured and perpetuated through the establishment of exclusivity, of limiting access to the most desirable communities and the best schooling. The value of place is determined only partially by the physical structures and locational advantages it affords. It derives also

AUTHORS' NOTE: *Some of the work that provided the basis for this publication has been supported by funding under a cooperative agreement between the University of Pennsylvania and the U.S. Department of Housing and Community Development. The authors are solely responsible for the accuracy of the statements and interpretations contained in this paper and such interpretations do not necessarily reflect the views of the government.*

from the characteristics of the others occupying the same space or the same schools, and is financially expressed in a class monopoly rent (Harvey, 1974), essentially the return on contrived scarcity. Recent American trends toward gentrification on the European model indicate that no iron law of location prevails, whereby rich people live on the geographic periphery and the poor reside at the center. Rather, social center and periphery are determined by where rich and poor live and work. The wealthiest choose between rehabilitated Victorian and spacious ranchstyle for their domestic locus, then impart value to the surrounding properties. Poor people fill in the gaps once the exits and loyalties of the well-to-do and their middle class emulators take shape.[1]

Within this dynamic of locational choice, the consequences of economic change and governmental actions that affect urban neighborhoods are not important for their effects on place per se, but on social space; not for the benefits captured by a particular area or the condition of the buildings within it, but for who stays there and who leaves. Our aim in this chapter is to examine the tension between mobility and community as it plays itself out within the urban neighborhood and is affected by public policy. In particular, we are concerned with the use of citizen participation as a device for mediating tension and its theoretical potential for enhancing communal solidarity, especially in the interests of low-income groups. We approach this task first by examining the macro factors which define neighborhood dynamics and determine the effects of participation; second, we consider the present policy situation as reflected in the Community Development Block Grant Program and other federal programs; finally, we ask whether neighborhood activity, including but not limited to official citizen participation, can change the macro-situation of lower-income households at the neighborhood, city, or national levels.

NEIGHBORHOOD DYNAMICS AND PUBLIC POLICY

The individual choice to stay or move results from dynamics of place and person. Even without a major change in a household's income or family situation, neighborhood decline or improvement reflected in increased rents may cause a family to move. Private control of housing markets means that pricing decisions in a neighborhood cannot be determined by public policy, although they are, of course, influenced by it. Both upward and downward movements in prices destabilize neighborhoods and divide communities according to the disparate economic interests of their residents. Residential populations become hostage to the constant search for new investment opportunities, and poor people especially are unable to shield themselves from the changes in property values produced by shifting investments. New investment is rationalized as being in the public interest through the dynamics of filtering. However, filtering, even if it worked as well in practice as in theory, promises the poor improved housing quality through mobility rather than community improvement.

Private suburban construction as well as large local subsidies through tax abatement and publicly backed bonds to downtown development contribute to neighborhood turnover. Suburban construction drains those central cities caught in downward spirals of disinvestment; the spillover of downtown development in cities such as San Francisco and Denver increases the value of centrally located land and forces out low- and moderate-income residents. For individual members of the working class, suburban residence provides the appurtenances of a middle-class style of life. At least until recently, it has signified a genuine improvement in living standards for millions of people. But it is an improvement realized individually rather than collectively, and through its separation of home from work, white from black, high wage from low wage, owner from renter, it has divided the working class and limited its collective political potential.

The federal government has intervened in local housing markets with the intention of providing a decent home and neighborhood for everyone. In the past it offered incentives to suburban outmigration through mortgage guarantees and caused people to move from central-city locations in the name of slum clearance. It continues to encourage geographic mobility notwithstanding the preservationist aims of the present governing statute and the chorus of criticism leveled at previous urban renewal legislation for its dislocating effects (see, inter alia, Gans, 1962; Fried, 1966; Hartman, 1979). Confusion prevails, however, over whether government efforts are to be evaluated according to their impact on low-income citizens or their effect on the physical structures in which those citizens once lived (see Downs, 1979). One of the aims of the Housing and Community Development Act (HCDA) of 1974 is "the reduction of the isolation of income groups within communities . . . and the revitalization of deteriorating or deteriorated neighborhoods to attract persons of higher income" (Title I, Sec. [c] [6]). This goal implies the improvement of place through outmigration of the poor rather than through amelioration of their condition in their present locales. While HCDA also stresses individual rather than areal benefits under the Section 8 rent subsidy program, here again the administration's intent is to promote mobility of low-income households. Moreover, HUD's prohibition on new Section 8 residential construction in racially homogeneous areas undermines population stability by precluding increases in the supply of housing in neighborhoods where low-income minority people may have lived most of their lives and where housing shortages have resulted from demolition and abandonment.

At the same time, HUD emphasizes the importance of the commitment of current residents to their neighborhoods. "Many community development activities depend upon private investments and specific actions by individuals fixing up their homes and by neighbors working together to improve their communities" (USDHUD, 1978a: 1). Within the confluence of private investment and public policy joining to produce constant flux in the character of urban districts, citizen participation is being advertised as a force to keep the agents of change under control. According to Title VII in the 1978 amendments to HCDA (Sec. 702 [a], entitled the Neighborhood Self-Help Development Act),

> (1) existing urban neighborhoods are a national resource to be conserved and revitalized wherever possible . . . ; (2) to be effective, neighborhood conservation and revitalization efforts must involve fullest possible support and participation of those most directly

affected at the neighborhood level; and (3) an effective way to obtain such support and participation at the neighborhood level is through neighborhood organizations accountable to residents of a particular neighborhood with a demonstrable capacity for developing, assisting, and carrying out projects for neighborhood conservation and revitalization.

By this reasoning, stable, socially integrated neighborhoods become the result of organized citizens shaping programs in conformity with their common interests. However, community groups must often confront faceless economic and political forces, many of which operate differentially among the members of the community, posing a constant threat to neighborhood cohesion. The instability and political weakness of community groups is magnified by the relationship between social and geographic mobility, which means that community objectives are frequently at odds with individual self-interest and the thrust of governmental policy.

THE COMMUNITY DEVELOPMENT BLOCK GRANT PROGRAM (CDBG)

The most important federal effort now aimed specifically at improving neighborhoods is the Community Development Block Grant Program (CDBG). The allocative process within cities under this program has become a principal arena wherein neighborhood interests seek to regulate urban change. The outcomes of this process significantly affect the stability of neighborhoods. It is therefore useful to examine briefly the conflicting objectives of CDBG, the place of citizen participation in the program, and the implications of federal programs for the local politics of community development.

Congress established the CDBG Program through Title I of the Housing and Community Development Act of 1974 (P.L. 93-383). CDBG was one of several programs in the "new federalism" of special revenue-sharing and increased authority for local governments. It supplanted seven previous categorical programs, the most significant of which were Urban Renewal and Model Cities. CD funds were to be distributed on a formula rather than a project basis, with the bulk going to approximately 540 entitlement cities larger than 50,000 in population.

Within broad guidelines, the cities could use CD money for capital improvements and housing rehabilitation, but with the important restriction that the program could not finance new construction. The act was extended for three more years in 1977 (P.L. 95-128) and modified by the introduction of a second formula and strengthened language requiring targeting of benefits to lower-income areas. By fiscal year 1979 CDBG was budgeted at somewhat more than $4 billion annually.

The 1974 Act contains language which reflects conflicting interpretations of the objectives of community development—whether the program should be aimed at people or places, whether the "improvement" of a neighborhood means the replacement of low-income populations by more affluent residents, whether homogeneous minority and poor communities should be conserved or their populations dispersed. The legislation states that its primary objective is "the development of viable urban communities, by providing decent housing and a suitable living environment and expanding economic opportunities, principally for persons of low and moderate income" (P.L. 93-383, Sec. 101 [c]). Yet, immediately thereafter, a subsidiary objective becomes "the elimination of slums and blight," a phrase often interpreted to imply urban renewal and land clearance. Title I specifies the aim of "conservation and improvement of the Nation's housing stock," again for the principal benefit of lower-income groups. However, CDBG offers no new low-income housing, no benefits to renters (who constitute a majority of the poor), and no guarantee that occupants will not be displaced by higher rents from "upgraded" housing. While CDBG also claims as an objective "the expansion and improvement of the quantity and quality of community services," HUD has discouraged service expenditures by imposing a 20 percent ceiling, and most cities have chosen to devote even smaller proportions of their budgets to this end. Thus, the only element in the program which specifically distributes benefits to low-income people receives little emphasis.

The 1977 revisions and the ensuing regulations tightened the language relating to beneficiaries by changing the word "or" to "and" in a section specifying criteria for HUD review of local programs: A city must give "maximum feasible priority to activities which will benefit low *and* moderate income families or aid in the prevention or elimination of slums and blight" (Sec. 104 [b] [2] as modified, emphasis added). Although the alternative goal of "eliminating blight" remained in the statute, HUD further emphasized the benefit criterion in its 1978 regulations, such that the burden of proof lay with an entitlement city unless it targeted 75 percent of its program toward "lower income" neighborhoods.

In practice, an increasing share of benefits have been going to such neighborhoods since the initial years of the program. However, city CDBG programs themselves incorporate the divergent objectives of the act, so that there is no dominant pattern of local commitment to stabilizing lower-income communities. Street improvements and housing rehabilitation for homeowners in "transitional" neighborhoods meet the HUD definition of "maximum feasible priority." They also may result in enhanced property values, displacement, and gentrification. Thus, while CDBG is a great improvement over Urban Renewal, it is a step backward from the Model Cities goal of improving the physical condition of lower-class neighborhoods and the social welfare of their lower-class occupants.

THE POLITICS OF CITIZEN PARTICIPATION IN CDBG

CDBG constitutes the main vehicle through which lower-income residents may participate in planning and development activities affecting their neighborhoods. Citizens and organizations are granted the opportunity to express their concerns at all stages of city-run programs. In some places, they perform significant planning functions, and, in a few, they implement local programs.

The architects of CDBG, however, designed the program so as to avoid the political liabilities of two previous efforts to revitalize inner-city neighborhoods: the Community Action Program (CAP) and Model Cities. Each of these programs established local administrative agencies which were at least semi-autonomous from city governments, with independent technical staffs and relatively strong citizen participation requirements, including the representation of target-area residents on agency governing boards. In contrast, local CDBG programs are directly controlled by mayoral agencies; the federal government deals only with city officials; and, under CDBG, cities are prohibited from delegating any program authority to independent agencies, whether citizen-controlled or not.

Citizen participation requirements were somewhat strengthened in the 1977 statute and 1978 regulations, but they remain weak and vague in comparison to the programs of the sixties. Citizens are given only advisory power over city programs. The cities must hold public hearings, respond to citizen complaints, and write a citizen participation plan (P.L. 95-128, Sec. 104 [a] [6]). However, there are no federal specifications of what must be contained in the plan. In fact, HUD only requires certification by city officials that a plan exists, rather than HUD review of its content. Given these loose federal regulations, it is not surprising that cities have structured citizen participation in a variety of ways, and that the extent of neighborhoods' influence, while nowhere great, depends upon the balance of local political forces.

In our own research we are endeavoring to assess the effect of community group activity on CDBG outcomes as part of an HUD-sponsored study, centered at the University of Pennsylvania, of CDBG in nine cities. These are New Haven, Pittsburgh, St. Paul, Memphis, Birmingham, Corpus Christi, Wichita, Denver, and San Francisco. The cities represent a considerable range of political contexts and histories of group and neighborhood activism. While the overall research project—entitled "Community Development Strategies Evaluation"—is concerned with many aspects of the neighborhood and household impact of CDBG activities, our main focus is on neighborhood organizations, citizen participation, and, more generally, the local politics of community development. Our approach is to have field researchers spend several weeks at a time in each city, conducting semistructured interviews; observing meetings; collecting budgetary and other data; and developing case studies of organizations, projects, and decisions. We have been in the field at this writing for about nine months. Based on this limited period of observation, we can present some tentative generalizations concerning the impact of citizen participation on community development.

We find that much community participation in all the study cities moves through the traditional modes of interest group pluralism and the lobbying of city officials, rather than being limited to the formal participatory mechanisms established under the program. CDBG is thus similar to any other capital expenditure program in the kind of politics it creates. Within the framework of pressure and bargaining which is implied by normal urban political decision-making, initial resources of participants count a great deal. As a result, overall capital expenditures of the cities, as opposed to those which are specifically funded by CDBG, reflect the traditional redevelopment constituency of business, especially real estate and tourist industry interests. This is particularly true of major capital construction projects funded by EDA, UDAG, and municipal bonds, which have important spillover effects but over which community groups have virtually no say. The part of the capital budget which is attributable directly to CDBG is more responsive to community group input, but not necessarily through the formal citizen participation as opposed to the informal lobbying process. Some neighborhoods, commonly those which once had a Model Cities program, stand out as able to influence the resource allocation process.

CDBG tends to be a mayor's program in cities where there is already a strong mayor. In all cases, however, the critical decisions concerning CDBG are made centrally. Despite the rhetorical stress on neighborhood which surrounds CDBG, it is a neighborhood program only in the sense that most expenditures attributed to it go to residential areas. Despite the increased neighborhood emphasis arising from the 1978 regulation calling for the designation of neighborhood strategy areas, there is less of a neighborhood focus than in the preceding programs of Community Action and Model Cities, both of which concentrated exclusively on a few target areas, or only one, in each city. By the time a CDBG funding allocation reaches a neighborhood, the influence of neighborhood groups on its use is marginal, since its purpose is already designated. For neighborhood residents to affect the program, they must intervene in the decision process at an earlier point.

While the switch from categorical to block grant embodied in CDBG has increased the flexibility of the city government in addressing development issues, it has, in various ways, limited the leeway of community groups. First, community groups cannot use the power of the federal government as a resource when pressing demands against the city. Second, while the program provides funds for activities to which city planning departments accord first priority (infrastructure in particular), it constrains severely the amount of attention that can be given to certain functions with high priority to low-income citizens. The restriction, which in some cities is interpreted by the CD office as an interdiction, on social service spending means that job-creating and service-receiving opportunities for low-income people are precluded. While CDBG does finance construction jobs, these are of less interest to neighborhood people than the paraprofessional positions of Community Action, Model Cities, and, in some places, the Comprehensive Employment and Training Act (CETA).

Third, for those low- and moderate-income individuals who are not homeowners, CDBG offers few prospective rewards. For them the housing problem is felt most intensely as the lack of low-cost rental housing; the major CDBG emphasis on rehabilitation assistance to private owners creates a trade-off between improved housing quality and low rents. The objective of lower-income renters is the construction of affordable housing, but new construction is prohibited under CDBG. Assistance to low-income housing sponsors is the only use of CDBG funds which really addresses their concerns, but only a miniscule proportion of funds has been used for such purposes. Thus, rent levels and housing supply remain as givens; participation in the CDBG allocation process for the most part is not an effective means to counter the upward pressure on rents which results from neighborhood revitalization. Given the extremely limited nature of Section 8 rent subsidies and the restrictions on the uses of CDBG, there are few good strategies for renters to follow. The problem is further exacerbated within minority neighborhoods, where HUD prohibits any new subsidized construction because of racial impact.

Effective community group participation depends on active local leadership, coalition-building, and the ability to gain technical assistance. Differences among neighborhoods in terms of their ability to gain a hearing can be traced almost entirely to the effectiveness of their representatives. Only through coalitions are community groups able to influence overall priorities of the CDBG program and avoid the destructive competition which gives city officials a free hand. The capital-intensive nature of the program and the domination of the agenda-setting phase of the decision process by professional staff mean that community groups are at a serious disadvantage, especially in terms of program evaluation, if they do not have access to technical assistance. HUD has refused to deal directly with community groups, and thus does not constitute a source of such help. Community groups have looked to private foundations, other federal agencies, or federally sponsored groups like Neighborhood Legal Services, VISTA, and the Center for Community Change for aid. They also receive assistance from the city itself, but it is rare that city-funded planners will oppose the administration.

Unless City Hall genuinely wishes to encourage citizen participation, the CDBG program will not precipitate it. This is true both because CDBG does not provoke the extreme opposition of urban renewal and urban highways and because it does not provide substantial support for citizen participation. In some cities the mayor considers community groups to be a useful element in the governing coalition; in others he or she regards them as a threat. Planning staff have become more sensitive than previously to the need for neighborhood preservation and the impact of community development programs on low-income groups. Nevertheless, cities continue to put more resources into economic development than neighborhood improvement programs. The inadequacy of funds to assist low-income people was everywhere lamented by planners we interviewed, but this was not reflected in a diversion of funds away from major capital improvements into leveraging investment in low-income housing. The deference paid in words to the needs of low-income groups has not been translated into funding comparable to that given to waterfront development, commercial revitalization, new hotels, and the like.

Conscious social policy, as defined by programs administered by the former Department of Health, Education and Welfare (HEW), has, like the HUD programs, simultaneously incorporated emphasis on place and people. Community action and community mental health programs have stressed the situational determinants of poverty and have attempted to change the neighborhood social environment. Of far greater magnitude, however, are the income transfer, education, and training programs directed at individuals. The criteria for success of these latter programs have always been measured in terms of individual mobility—getting off welfare, graduating from high school, finding a job. With the demise of the community action programs, class and group interests are mainly addressed through affirmative action programs and other efforts to eliminate discrimination. The value of competitive individualism underlies social policy. Community has never been a strong legitimating formula within the United States (contrast the slogan "life, liberty, and the pursuit of happiness" with "liberty, equality, fraternity"). The working-class quarter, where interest was overlaid with affect, existed at only brief moments in American history. In Louis Hartz's (1955: chap. 8) phrase, "Algerism" became the way by which industrial America justified within a liberal-democratic ideology the creation and maintenance of inequality.

Politics, within this framework, becomes the contest of individuals or special interests rather than classes. Attempts to create political solidarity on the foundation of a common territorial interest like the urban neighborhood are doomed to founder when leaders confront material threats or incentives. Preservation and stabilization constitute discordant themes in the American litany. The dominating value within American ideology has been mobility, both social and geographic, as illustrated in this quote from *Street Corner Society:*

> The history of the [Italian Community] Club demonstrated that its two aims, the social advancement of the members and the improve-

> ment of conditions in Cornerville, could not be realized by the same people at the same time. The college boys were primarily interested in social advancement. The corner boys were primarily interested in their local community [Whyte, 1943: 97].

Moving up has traditionally meant moving out, although these days it may mean moving back. The new emphasis on neighborhood contradicts the old values of opportunity and motion, privatism and competition. Few "rational" Americans would sacrifice individual success to a vaguely defined communal interest, especially when individual achievement has always been touted as an indicator of collective welfare. According to the American ideology, doing well is doing good.

Neighborhoods therefore are weak bases on which to build a structure to represent lower-class and working-class interests. Even when the political machine operated from a highly organized neighborhood constituency, it distributed specific rather than communal benefits. There is now a popular ideology which supports neighborhood organization for collective goods (see Fainstein and Fainstein, 1974: chap. 1.; Fainstein and Martin, 1978), but community groups with such aims must operate in an institutional context largely unsupportive of their ends. Progressive reform, when it destroyed the machine's partisan organizational basis, severed the link between neighborhood activity and central decision-making. Neighborhood interests, as we noted earlier in our discussion of CDBG, must approach decision makers through lobbying and pressure; they are outsiders rather than normal components of the allocative process. As a consequence, neighborhood participation in city politics tends to pit one district against another rather than all against the center.

The present thrust toward community participation operates within two divergent normative frameworks, that of community activists who regard city government antagonistically, and that of governmental policy makers who see community organization as a potential vehicle for legitimization and policy implementation. These two views have existed in tension since the original provisions for "maximum feasible participation" under the Economic Opportunity Act of 1964 (see Rubin, 1969;

Fainstein and Fainstein, 1974: chap. 2; Moynihan, 1969; Cloward and Piven, 1972). Considerable subsequent debate has not settled the issue, either among regime supporters or left critics, as to whether the government's participatory mandate was a cooptive response to the threat brought by mobilized minority groups or an incentive to unrealistic expectations and unconstructive protest, mistakenly proferred to militants. The simultaneous diminution of both strong governmental mandates for participation and minority mobilization makes retrospective disentanglement of cause and effect difficult.

The legacies of the community activism of the late sixties and early seventies are a less vocal but nevertheless substantial network of community organizations and the development of a cadre of neighborhood leaders (see Lancourt, 1979: chap. 7). On the government side, the weakening of the federal requirements for community participation has been accompanied by a proliferation of programs, estimated now at 226 (National Commission on Neighborhoods, 1979: 279), that require some form of citizen consultation. The concluding section of this chapter examines whether this infrastructure of organizations and programs is likely to have a significant influence on neighborhood quality and stability.

NEIGHBORHOOD ACTIVISM AS A POLITICAL FORCE

Organizations and neighborhood groups which are involved in formal participatory structures constitute only a part of the large number of groups now to be found in most American cities. Political officials, academics, and activists have, in recent years, identified an important—some say rapidly expanding—neighborhood movement. It is possible that the neighborhood movement can accomplish objectives which we see as unlikely to be achieved through official citizen participation. Does this movement—a term we will accept as a matter of convenience—substitute community and collective social mobility for individual exit and gentrification? In order to assess the neighborhood movement and its potential we need to consider (1) its character, strength, social bases, and objectives; and (2) the potential efficacy of that part of the movement most committed to the communal objectives of lower-income neighborhoods.

In the waning months of a decade of fiscal conservatism and retrenchment from the social goals of the sixties, claims were advanced for a neighborhood movement of major political significance (Boyte, 1979). There is, however, little consensus as to the political character of this movement, its internal homogeneity, and the extent to which the movement as a whole reflects the interest of lower-income and minority neighborhoods. In February 1979 the journal *Social Policy* sponsored a National Conference on Neighborhood Organization, with funds from ACTION. Participating in the conference were both scholars and neighborhood leaders. The conference manifested large differences in perspective on the neighborhood movement, as the following quotations suggest.

> The neighborhood movement of the 1970's represents an emerging social force which has the potential to bring new meaning to our concepts of citizenship and democracy.... Having arisen precisely because of the failures of both representative democracy and government mandated citizen participation to reflect the needs of low- and moderate-income people, the neighborhood groups are striving to make existing institutions more accountable and to gain increased control over the decisions that affect their lives [Perlman, 1979: 16].

> As the administrative and organizational side of the neighborhood movement has grown stronger, its political side has grown correspondingly weaker (at least as an independent voice) [Mollenkopf, 1979: 26].

> There is a popular misconception that neighborhood organizations across the country, paraded popularly sometimes as the "neighborhood" movement, make up a primarily lower-class and minority movement. That is not the case.... White, middle-class neighborhoods are better organized and better able to influence decisions affecting them than are poorer neighborhoods [Jones, 1979: 45].

It is not surprising that there should be so much disagreement about the neighborhood movement. With more than 4000 organizations and groups in several hundred large cities, different scholars and activists may be describing different samples of groups, or simply different political contexts. In addition, many scholars have political commitments to their version of the movement and inevitably look for examples of success, particularly since organizational failures by their very nature disappear. Thus, one should be cautious about making strong assertions. At best, we can make a few generalizations about which there might be consensus:

(1) The movement is rooted at the local level, and is only slowly organizing into national coalitions. It is not very influential as a national urban lobbying group.
(2) In any city, neighborhood organizations differ in size, structure, objectives, and social composition; there probably are more organizations which are primarily middle-rather than lower- or working-class in composition.
(3) For the most part, there is little connection between the neighborhood movement and trade union organizations.
(4) Lower- and middle-income neighborhood organizations have divergent objectives. Lower-income people want housing, jobs, and other material resources. Middle-income people are interested in beautification, infrastructure, and protection from external forces (which sometimes include lower-income people). Most organizations, regardless of class, want more participatory democracy and restriction of business influence on urban government. Neighborhood organization and citizen activism usually create conflict between the residential interests of neighborhoods and the business interests of "downtown."
(5) Lower-income organizations and those rooted in the middle class tend to function in separate spheres. The most common conflicts within lower-income areas have involved competition for program funds. The most important conflicts between organizations and neighborhoods along class lines have been associated with racial integration, especially school busing.
(6) The neighborhood movement is not and cannot substitute for local and federal commitment to and investment in lower-income communities.

The neighborhood movement has been verbally applauded by federal officials. This official support usually ignores the notion of inherent conflict between "neighborhood" and business interests, as when President Carter says "we must have a partnership--between a government which knows its limits, a private sector which is encouraged to do the right thing, and the people in their families and neighborhood and voluntary organizations" (quoted in Perlman, 1979: 19). The neighborhood movement is used to legitimize the system. The ability of residents to organize demonstrates the strength of grassroots urban democracy, just as the downfall of the Watergate criminals was heralded as a sign of democratic vitality in American institutions. While much of federal urban policy, as embodied in UDAG and EDA, facilitates business accumulation, the neighborhoods, officials suggest, have voluntary activism, which may well be more effective than federal money anyway. Thus, the Assistant Secretary of HUD for Neighborhoods, Voluntary Action and Consumer Affairs, Monsignor Geno Baroni, said: "New York [city government] . . . just can't get it together. *Twenty neighborhood groups have accomplished more in recent months than all of the federal programs*" (quoted in Herbers, 1970; emphasis added).

The second question at hand is what can be accomplished by that part of the neighborhood movement rooted in lower-income communities. Here we must distinguish between possible movement effects on three levels of political organization—communal or neighborhood, citywide, and national. Neighborhood organizations have engaged in a wide range of activities, primarily at the communal level. However, precisely because of the localism of organizational bases and objectives and their relatively weak resources in the face of governmental and business institutions, even successful communal action rarely produces a significant change in the forces which make all neighborhoods—especially those with lower-income and minority populations—dependent on external economic and political factors. High levels of participation and effective organization have contributed to neighborhood stabilization in at least two of our study cities. Milton Kotler (1979), Executive Director of the National Association of Neighborhoods, summarized research showing many more instances of similar success on a wide range of fronts, from improving service delivery to changing zoning decisions and blocking highway projects.

Yet, we can also point to many instances where effective organizations, having attracted investment to their neighborhoods, could not then stem the ensuing tide of gentrification—to the point where their own membership was converted to middle-class homeowners, and their goals redirected to "historic preservation." Similarly, organizations may be able only to retard a disinvestment spiral, as in the case of one of the most long-lived and purportedly effective groups, The Woodlawn Organization (TWO), founded in 1960 by Saul Alinsky's Industrial Areas Foundation. TWO could do no more than slow the death of its neighborhood; that, however, may represent the best that can be expected of the neighborhood movement in our older cities. As a recent study reported, "instead of holding TWO responsible for not reversing neighborhood deterioration, there is a strong feeling among leaders and observers that the organization is responsible for preventing the complete removal of Woodlawn" (Lancourt, 1979: 166).

In general, the urban political movements of the sixties and the neighborhood movement of the seventies have proved most effective when vetoing external intrusions, which sometimes include minority residents and school children in white working-class neighborhoods. Attracting investment, and then controlling its outcomes so as to avoid displacement and gentrification, is very difficult to accomplish at the communal level. The appearance of community housing organizations—among other places, in New York, Denver, and San Francisco—represents an effort to build affordable housing for present residents; yet here, too, the local possibilities are severely limited by the downtown and middle-class orientations of city governments and the failure of the federal government to provide housing funds sufficient to effect more than a trivial impact.

The basic planning and expenditure decisions which determine the fates of neighborhoods are made at the city level by both public officials and market actors. It is the former who are the main target of coalitions of neighborhood groups so that they will channel the activities of the latter. Here the crucial objective is for lower-income activists and organizations to become part of the governing coalition, to make the regime receptive.[2]

Citizen participation in itself is not an effective means for reorienting city regimes. The final report of the National Commission on Neighborhoods (1979: 282) stated that "the Commission's case studies failed to reveal groups which used citizen participation mechanisms effectively, unless the group had already developed power through independent action." Such action necessarily means either militancy or the mobilization of electoral power, with the electoral route by far the more common choice in recent years. In some cities some neighborhoods seem to have joined the governing coalition. When that happens, city officials seek to accommodate both neighborhood and business interests or to channel investment in acceptable ways—or to let weak neighborhoods pay the cost of the protection afforded to the better organized. But the city itself, of course, is dependent on market forces operating at regional and national levels, so that only federal action can change the basic trajectories of urban development.

At the federal level a neighborhood movement could create the most important economic resources for determining neighborhood development: publicly funded new low-income housing, social welfare programs, and tax incentives and expenditures. There is now a neighborhood presence in Washington where none existed before. Although an organization like the National Association of Neighborhoods may not be lobbying primarily for a lower-income constituency, at least it represents some alternative to the U.S. Conference of Mayors, which previously claimed to represent "city" interests. However, it is hardly reasonable to expect that the neighborhood movement—operating outside of the political party structure, itself divisible along class and racial lines—can establish a force sufficient to redirect national policy. Citizen participation, whether in official structures or in the larger neighborhood movement, will be unable to redefine the main choices now available to the majority of lower-income neighborhoods—further decline or gentrification. Mobility, rather than community, will continue to be the American way out of addressing the problems of lower-income neighborhoods.

NOTES

1. Albert O. Hirschman (1970) analyzed the response to decline in institutions through the concepts of exit, voice, and loyalty. Exit corresponds to reliance on mobility or escape; voice refers to communal action for change; and loyalty is the prerequisite to voice. He goes on to analyze the situations under which people will stay or go and the likely effect on institutions of which strategy is adopted by which classes of people. Fundamentally, he discussed the relationship between markets (exit) and politics (voice) as controlling mechanisms.

2. See Browning et al. (forthcoming), who describe the importance of "regime receptivity" in determining city expenditure patterns.

REFERENCES

BOYTE, H. C. (1979) "Neighborhood power—a term representing a new constituency entering national political life." New York Times, August 19.

BROWNING, R. P., D. R. MARSHALL, and D. H. TABB (forthcoming) "Implementation and political change: sources of local variations in federal social programs," in P. Sabatier and D. Mazmanian (eds.) Effective Policy Implementation. Lexington, MA: D. C. Heath.

CLOWARD, R. A. and F. F. PIVEN (1972) The Politics of Turmoil. New York: Pantheon.

DOWNS, A. (1979) "Key relationships between urban development and neighborhood change." Journal of the American Planning Association 45: 462-472.

FAINSTEIN, N. and S. FAINSTEIN (1974) Urban Political Movements. Englewood Cliffs, NJ: Prentice-Hall.

FAINSTEIN, N. and M. MARTIN (1978) "Support for community control among local urban elites." Urban Affairs Quarterly 13: 443-468.

FRIED, M. (1966) "Grieving for a lost home: psychological costs of relocation," pp. 359-379 J. Q. Wilson (ed.) Urban Renewal: The Record and the Controversy. Cambridge, MA: MIT Press.

GANS, H. (1962) The Urban Villagers. New York: Free Press.

HARTMAN, C. (1979) "Comment on neighborhood revitalization and displacement: a review of the evidence." Journal of the American Planning Association 45: 488-491.

HARTZ, L. (1955) The Liberal Tradition in America. New York: Harcourt Brace Jovanovich.

HARVEY, D. (1974) "Class monopoly rent, finance capital, and the urban revolution." Regional Studies 8: 239-255.

HERBERS, J. (1979) "Activist groups are becoming a new political force." New York Times, June 18.

HIRSCHMAN, A. (1970) Exit, Voice, and Loyalty. Cambridge, MA: Harvard University Press.

JONES, D. J. (1979) "Not in my community: the neighborhood movement and institutionalized racism." Social Policy 10: 44-46.

KOTLER, M. (1979) "A public policy for neighborhood and community organizations." Social Policy 10: 37-43.

LANCOURT, J. E. (1979) Confront or Concede. Lexington, MA: D. C. Heath.

MOLLENKOPF, J. (1979) "Neighborhood politics for the 1980's." Social Policy 10: 24-27.

MOYNIHAN, D. P. (1969) Maximum Feasible Misunderstanding. New York: Free Press.

National Commission on Neighborhoods (1979) People, building neighborhoods; final report to the President and the Congress of the United States. Washington: USGPO.

PERLMAN, J. (1979) "Grassroots empowerment and government response." Social Policy, 10: 16-21.

RUBIN, L. B. (1969) "Maximum feasible participation: the origins and implications." The Annals, September: 14-29.

USDHUD [U.S. Department of Housing and Urban Development] (1978a) Citizen Participation in the Community Development Block Grant Program: A Guidebook. Washington, DC: U.S. Government Printing Office.

——— (1978b) Community Development Block Grant Program: Third Annual Report. Washington, DC: U.S. Government Printing Office.

WHYTE, W. F. (1943) Street Corner Society. Chicago: University of Chicago Press.

Part V

Informing the Local Decision Maker

□ IN PARTS II AND III, we presented a strongly structured view of the nature of individual mobility and the dynamics of small-area change. The various authors suggest that we can construct models of this behavior and can derive, analytically, at least some of the main impacts of federal, state, and local programs. However, when we look at the way in which these ideas have been utilized at the city level in local government planning and policy decisions, we find the influence is minimal. Why is this, and how might the situation be improved—at least to the extent of making those in the research and public policy arenas more aware of each others' activities?

One major source of difficulty in absorbing research outputs into the policy process is the importance of political agenda in institutional decision-making. This point is developed by two representatives of local governments. Martin Goldsmith and James Lemonides indicate that the nature of the policy-making process at the city level, with its emphasis on short-run problem-solving against a backdrop of embedded interests and constituencies, stands in stark contrast to the search for that general understanding of urban process which lies at the core of both academic research and federally sponsored evaluation studies. While recognizing the pertinence of information about mobility and change, the main pressure is to support existing policy rather than to incorporate new knowledge, since the latter almost invariably requires reevaluation and a change in policy direction.

James Hartling then goes on to suggest that academic research has responsibilities to assist the local decision maker as well as work in the more traditional arena of national and regional concerns. Specifically, he argues that researchers should be involved in assisting local practitioners in allocating resources. This can be achieved by a greater concern for policy mechanics and that the specific public actions remain consistent both with national goals and with the understanding of social process obtained from prior analyses.

It is at the local level that the greatest difficulties are encountered in translating general research findings into statements supportive of particular decisions and in communicating this translation in a politically effective way. Existing analytical studies tend to present two types of problems: lack of detailed context and narrowness of focus. For example, Weinberg (Chapter 10) suggested that outcomes of housing allowance programs might differ as a function of local housing market conditions. Attempts to estimate the potential effects of such programs in other cities should then require that these local market circumstances be identified. Yet, little effort has been devoted to specifying the appropriate indicators of market context, such as the characteristics of the existing stock, the composition and rates of growth of the population, and the range of current housing and related programs. Furthermore, even when we are able to calibrate our models for a variety of cities, their usefulness is greatly limited by their focus on problems which are analytically tractable. The fact that the aged, the unemployed, and the two-wage-earner family do not fit readily into our standard paradigms of locational choice is of little help to those responsible for implementing programs for these groups. Thus, as Goldsmith observes, accusations are made that research typically produces models which are too sophisticated for local needs while, at the same time, ignoring many of the significant elements of the problems they are supposed to be addressing.

Often, the primary need at the local level is for a better understanding of the social, economic, and demographic changes taking place within the community. All too frequently, however, the only accessible source of data is the prior census which, by the end of the decade, is hopelessly out of date. The ability to provide up-to-date monitoring of changing local conditions is invaluable to many decision-support situations, yet the development of such a capability is not merely a question of obtaining sufficient financial resources. Janet Byler and Randolf Gschwind take the position that while a great deal of potentially useful data are already collected in the course of administrative activity, the organization of such data into a monitoring system requires a great deal of thought if the policy process is to be effectively informed. Their chapter both provides a motivation for the development of integrated micro-level data systems and suggests how the organization of such data systems for planning purposes might proceed from a technical standpoint.

In a very real sense, however, the Byler-Gschwind paper begs the question of whether the unstructured nature of local policy-making is capable of being informed by the proposed technology. The point was raised, particularly in the comments by Ralph Ginsberg, that the informational needs of different organizations involved in planning and policy are by no means compatible or equally well served by a highly structured, technically oriented data base. This issue raises many more questions than it provides answers. However, it should form a major focus for those social scientists wishing to contribute to local policy: What is demanded is that we examine in much more explicit fashion what Hartling calls "policy mechanics." We must explore the ways in which different institutions are informed by our understanding of local dynamics and the ways in which such understanding can be translated into specific public actions.

14

Academic Research and Public Policy Formulation

MARTIN E. GOLDSMITH and JAMES S. LEMONIDES

□ AS STAFF MEMBERS in the planning department of a major city with overall responsibility for Community Development Block Grant planning, our concern is for urban problem-solving in general and for the revitalization of the city's residential communities in particular. The nature and consequences of relocation and neighborhood changes, therefore, are of considerable importance to us; and one might reasonably expect that studies of mobility would have a significant bearing on our efforts as practitioners. Unfortunately, from our perspective the contribution of analyses of mobility to the formulation of public policy, at least at the level of local government, appears to be minimal. The purpose of this chapter, then, is to explore some of the reasons underlying this failure. In so doing, we will be concerned not only with the rather specialized area of mobility analysis, but more generally with the relationship between the academic producer and the governmental consumer, since the difficulty we encounter in this area appears to generalize to many other spheres of academic-political interaction.

DIFFICULTIES OF ACADEMIA-GOVERNMENT COLLABORATION

What emerges at the outset is that the inability of government and academia to effectively collaborate continues to arise from mutual suspicion and mistrust, a product of the long-standing and powerful tradition that universities and politics should not mix. From government's perspective, the insularity of "academics" leads them, at best, to a hopelessly naive approach to problems of the "real world"; at worst, it breeds (spontaneously, it sometimes appears to us) antigovernmental bias. For their part, universities view anything political as, ipso facto, corrupting.

Moreover, our impression, and that of many of our colleagues, is that this mutual antipathy is much stronger at the local (that is, city) level rather than at the state or federal levels of government. This may be simply a function of geographical proximity, but it seems also to be due to the similarity of requirements among major departments of the federal bureaucracy–particularly those agencies involved with research and development–and those of university departments. We tend not to seek out universities; we suspect their products both written and human. Graduate students frequently view government, especially local government, as the employer of last resort, while the research efforts of universities are rarely offered to–as opposed to intended for–local political decision makers or planners. The same behavior extends to the relationship between local government and major, quasi-public research foundations. Local government has had the consistent experience of either not being funded by such institutions or funded in a subordinate role with the primary research direction provided by an academic institution.

Our second major observation is that decision makers are under too much pressure or have too little time to absorb, much less to utilize, the research of academics. In part this is because such research is frequently ambiguous, hypothetical, long range, and difficult to understand. The counterargument is that policy makers are looking for immediate and simplistic answers to problems which are highly complex and for which only tentative answers can be supplied. To some extent, these comments form a part of the same, almost reflexive, tendency toward mutual caricature. In fact, the nature of the political process and the institutional styles of both government and academia mitigate strongly against the use of academic research in forming or reshaping urban policy. Our worlds run on different clocks with little concern for the long run, merely a series of short runs; we act out of different metabolisms and respond to different external pressures. In short, we exhibit most of the characteristics of systems concerned primarily with maintaining their own equilibria.

Above all, our purposes differ. Social scientists are concerned with an improved understanding of the phenomena they examine; the objective of research in so far as it concerns political decision makers and planners is the immediate release of techniques for problem-solving. It is, then, the different institutional goals which determine the content of research and policy formulation as much as they do the process by which academia and government engage each other or fail to do so. In order to understand this relationship better, we examine the nature of urban policy formulation and its relationship to research.

HOW URBAN POLICY IS MADE

There is no question that urban policy formulation relies but minimally on any coherent body of social science knowledge. The enactment of urban legislation takes place in response to (1) an amalgam of broad constituencies often deeply rooted in past relations, (2) somewhat abstract philosophies, (3) particular self-interest groups, and (4) to perceived crises. Thus, strategies for the revitalization of cities under the Great Society programs of Kennedy and Johnson held that massive public sector assistance in the form of categorical programs was the best way to alleviate central-city deterioration, to help the poor and other urban minorities. The New Federalism of their successors reflected an equally traditional belief that stimulating the private sector and passing monies through to local government in the form of flexible block grant programs would achieve essentially the same ends. And to the extent that it can be defined at all, current urban policy straddles the philosophical underpinnings of the previous Republican and Democratic administrations, but in any event is more responsive to national inflationary considerations and taxpayer revolts than to traditional constituencies and party philosophy.

As with most aspects of domestic policy, the situation is further complicated by the difference between legislative intent and what finally emerges as a program through regulations promulgated by the executive branch. These agencies remain substantially intact throughout changes of administration, and their ability to bend urban policy in response to their constituencies remains formidable in the face of major initiatives by the legislative branch. Thus, the Housing and Community Development Act was clearly intended to help local government broadly address the problems of prevention of residential blight as well as its elimination. However, successive regulatory requirements on the targeting of neighborhoods for concentrated activity have effectively minimized the preventive aspect of the legislation.

URBAN POLICY AT THE LOCAL LEVEL

The Brookings Institution has pointed out that activities at the local level have been carried out by agencies which operate with considerable independence of local officials, who, presumably, are ultimately responsible for local policy-making. That report further observes that the typical city council reflects the similarity of local and federal constituency-based initiatives, despite legislative intent by council members toward coordination. The block grant is viewed by council members as flexible money which can fund pet projects. A stream of demands continues throughout the program year in which the budget is continually shuffled around to free money.

The development of local urban policy thus reveals many of the same features and dynamics which characterize national urban policy formulation. It, too, is responsive to a host of embedded interests and constituencies; it is also sensitive to the interval between a proposed policy or program and a forthcoming election, although its policies undergo similar modifications by a bureaucracy which has its own survival to contend with and its own outside supporters. Chiefly, however, local urban policy has been dictated from above.

THE USES OF DATA, ANALYSIS, AND RESEARCH IN POLICY FORMULATION

The pervading issue in the above discussion of urban policy formulation is the pluralistic, constituency-based, short-term nature of intervention strategies proposed at all levels of government. The sheer momentum of these forces in shaping programmatic intervention is such that a major policy shift at the federal level is not likely to occur through the introduction of new knowledge on the urban condition.

The data used by practitioners are largely determined by the requirements of a given federal program, whether for purposes of determining initial eligibility and funding levels or in order to provide quantitative accountability measures to the appropriate federal department. Programs in operation need to be monitored for performance and compliance with regulation; other programs may be evaluated against certain standards to justify (or reject) recommendation of their continuance. Quantitative information is sometimes used as a basis for projection or description of needs for the application of programs, or the extension of existing programs. In all the above instances, data are used as feedback to redefine problems or to shift the emphasis of existing programmatic interventions.

The nature of the research undertaken by the practitioner closely parallels that of the formulation of urban policy itself: short-term, program-specific, and sensitive mainly to constituency. Research of a more academic nature which seeks to advance knowledge is missing. As a result, the quantitative information generated by this process has certain limitations. First, current data may not exist or may have only limited availability. The constraints imposed by a decennial census on documenting local conditions require no great elaboration; in 1979, local government used data which were a decade old. Congressional committees are thus frequently forced to fall back on incomplete, case study, or anecdotal data (usually furnished by consultants or interest-group testimony), a procedure which may have decidedly negative consequences. A good illustration of this point is the current spate of legislation designed to serve the disabled; these regulations require that government at all levels (and those who rely on its funding) engage in enormous capital outlays to serve a population whose needs have never been identified in detail. Indeed, some local government studies conducted after the fact would tend to indicate that much of the mandated response is out of all proportion, not only to the need but also to the demands of this population.

There is also the problem of time lag in the use of information; that is, there is a gap between the perception of new problems and the ability to shift policy in response to that perception. For example, urban programs are still locked into twenty-year-old strategies for combating urban blight and decay. However, many cities are experiencing neighborhood revitalization as a result of new forces, such as the structure of new household formation (see Hughes, Chapter 5 in this volume), concern for energy shortages, the emergence of new lifestyles, and women's rights. In short, many city communities are beginning to respond to an evolving perception that, after all, mature urban areas may hold considerable attraction with all that this implies for future policy.

As mentioned above, information and analyses required by government are intended less to generate new knowledge about existing conditions than to reinforce extant policy. In addition to suffering from the defects of being incomplete, outdated, and fragmentary, these data and the analyses or research which are derived from them exhibit further problems. Among these is the perhaps insurmountable difficulty involved in the construction of quantitative indicators to represent abstract concepts or phenomena which comprise urban problems. Regardless of the application, quantitative data are derived from static, point-in-time observations. Appropriate indicators are selected, along with observations for a particular "frozen moment," and the primary effects of chosen independent variables, which may or may not include policy instruments, on the dependent variable are observed. Secondary effects often are not considered, due to the additional complexity they would introduce. Much of the analytical work that begins with quantitative research invariably concludes with a more qualitative discussion of "externalities" which may have affected the results. Externalities, in somewhat circular fashion, are defined as any influence outside the descriptive scope of the research model, even though these contextual representations may be critical to an interpretation for policy purposes.

No better are the studies or derivative models based upon population or land use forecasting. Their application is suspect because consumers have more than a passing interest in the outcome of projections or forecasts; simply stated, the need for more "anything" usually means more federal dollars. Thus, predictive research inevitably yields optimistic results with respect to projected demand. Where there is a conflict of local interests as might arise in interjurisdictional disputes in which a metropolitan area and a city are competing for the same future population, more than one set of forecasts is used. Often there is nothing but political power to resolve the differences in future scenarios of population change.

Nor is predictive analysis particularly useful as a long-range policy or planning tool. For example, transportation policy illuminates most clearly the inherent limitation of predictive modeling, particularly as it has been extensively utilized by government. With its exclusive reliance on straight-line extrapolation of past demographic and income trends to predict car ownership and its inability to incorporate social and economic analysis, except in a secondary role, such policies seem infinitely better suited for generating demand than for predicting change. The consequence of the typical planning procedure has been to produce a significant contribution to (some would say "cause of") suburban expansion and urban decline.

Central to the descriptive shortcomings discussed above is the limitation of linear-causal approaches to an analysis of the dynamics of the urban condition. While, for example, linear least squares regression analysis recognizes that numerous factors may influence a particular condition, it does not consider the influence of factors upon each other (indeed, if two independent variables appear to be too closely correlated, one of them is dropped to increase the efficiency of the model). The definition of one (and only one) dependent variable per model necessitates an orientation toward simple linear causality, with all other variables perceived to affect only the dependent variable.

In summary, then, urban policy at all levels of government is determined by a variety of factors in which the role of data, analysis, and more extensive research on urban conditions is minimal. The nature of the data and the types of quantitative research which emerge are generally divorced from any consideration of broader context and appear to be incapable of describing, much less accounting for, the complex and interactive nature of urban communities.

15

The Public Policy Environment: Mobility Researchers' Responsibilities

JAMES E. HARTLING

□ I WOULD LIKE to focus my comments on a specific set of questions. First, what is the responsibility of the public sector to facilitate mobility? Second, to what extent does public policy reflect our knowledge of residential mobility? Third, what is the responsibility of the research community to promote its knowledge in the public sector? Fourth, what is the responsibility of public officials to seek the advice of the research community?

During the past two decades we have seen five different federal administrations; at least three different philosophies of urban development; three different approaches to federal-local relations; and a host of different federal programs for housing, community development, and economic development. A strong case can be made, however, that these many changes in public philosophy and programs have masked the durability of public policy intents. Our underlying views of the problems associated with or arising from residential mobility, the value judgments we make about those problems, and our views of appropriate public actions necessary to alleviate these problems have been constant for fifteen years or more. The current concern with "displacement" (or "recycling," as it is called in Philadelphia) can be viewed simply as a variation on a 25-year-old theme—namely, the response to the excesses of urban renewal and highway projects which resulted in the uniform relocation assistance and neighborhood development programs of the late 1950s and '60s. Furthermore, encouragement of dispersed housing opportunity for low-income and minority families and individuals has been a clear public policy objective for the past decade and a half.

As a public policy practitioner, I am struck that the intellectual and value content of public policy relating to residential mobility is enormously durable. Those research findings which identified problems associated with mobility and suggested remedies for them have had a long-lasting impact. It may be true that researchers have difficulty articulating the public policy implications of their work (a jaded practitioner might argue they do not care to articulate those policy implications) and even more difficulty imbedding those results in public policy. However, it also appears true that those results, once accepted in the public arena, endure as guides to public action for many years. As a result, we have an extensive range of values about residential mobility deeply imbedded in the public sector, each with an array of national programs intent on maintaining that value. There are in particular four very strong and durable public values about residential mobility.

(1) Resource limitations should not severely constrain access to home ownership by renters. This transfer is an important component of intraurban moves. When down payment requirements are too onerous, we respond with mortgage insurance. When interest rates are too high, we manage the money supply to lower them. When accessible land becomes scarce, we build federally supported highways. When shortages of building materials are feared, we move to avoid those shortages.

(2) Low-income families and individuals, as well as more affluent households, should be faced with a range of opportunities to which they could move to improve their housing. When the costs of housing preclude low-income occupancy, we provide reduced interest loans or direct rent subsidies. When zoning and subdivision restrictions preclude housing types

appropriate for low-income families and individuals, we attack those restrictions through the courts or withhold federal funds from those communities until those restrictions are lifted.

(3) Racial and ethnic minorities should also be presented with a range of housing opportunities. When these minorities also have low incomes, we utilize the programs intended to increase the mobility of low-income groups for this second purpose. When individuals discriminate against minorities in the sale or rental of property, we outlaw those practices and implement positive affirmative action in housing programs.

(4) Individuals and families should have the right to stay where they are if they so choose (or to receive substantial compensation in extreme cases when this is not possible). When urban renewal and public improvement projects relocate people, we provide uniform relocation assistance. When private-sector activities threaten to displace low-income persons, we respond with rental assistance and property tax deferral programs. When jobs move to other areas, we provide public improvements and business loans to produce new jobs in the distressed community.

It is clear, then, that the public sector has accepted very broad responsibility for facilitating housing adjustment through moving. It appears to me that the research community has played a significant role in encouraging this public sector commitment, largely because that community has directed its efforts to at least three specific tasks.

(1) The identification and analysis of problems associated with mobility, particularly the identification of mobility-impaired segments of the population. These problems are usually identified as national problems requiring national solutions, rather than as problems of a particular locality.

(2) The recommendation of public remedies for these problems, usually in the form of national legislation and programs. I should note that this approach reflects a laudable respect for economy of effort on the part of the research community, since national legislation appears, at least initially, to deliver the maximum impact.

(3) At a much later date, observation of the impact of these nationally mandated efforts. Very often, financial support for such observation comes from the federal agencies charged with implementing the programs. These agencies expect, as part of this observation effort, an evaluation of program implementation and recommendations for modification of day-to-day operating guidelines and procedures.

By focusing on this limited set of tasks, the research community has been quite effective in developing national policy commitments to access to opportunity with its implications for residential mobility. Yet, those of us charged with protecting these public values on a day-to-day basis by successfully implementing these many public-sector programs find little immediate help in the work of the research community. My job is to achieve these many objectives as efficiently as possible with limited resources. The findings of the research community suggest problems which should be ameliorated, regardless of the level of resources necessary. The research community approaches the issue of implementation almost solely in terms of national legislation. I deal with that legislation only as it is interpreted to me through regulations and guidelines. The intellectual content of those regulations and guidelines are seldom clear. I see the research community most often when it appears to conduct federally supported program evaluations. The most common results of such evaluations are changes in program reporting to the federal agencies. It is seldom clear how these modifications will improve program performance, although it is always clear how much extra staff time will be required to complete that reporting. My conclusion from this discussion is that the research community has an obligation to support the local government practitioner as he or she struggles to carry out the national commitments to residential mobility. There are in particular three responsibilities of the research community I want to highlight.

First, the research community has the responsibility to assist local practitioners to make short-run resource utilization deci-

sions. In the long run, when extensive resources have been allocated for the many programs spawned by our commitments to equal access to opportunity, the several objectives outlined above may all be achieved. However, in local government, our long run is really a series of short runs, and the resources available in each of these short runs are quite limited. Even this circumstance would not be too serious if these resources were useful for achieving only a single objective. But they are not. Often the same resources are essential for achieving alternate objectives involving mobility. In Philadelphia, for instance, we are engaged in a process of meeting with every white neighborhood in the city under a mandate to locate Section 8 rental housing widely throughout the city. That we are proceeding with this process despite the strenuous objections of many of these neighborhoods and their elected officials should not be surprising. It also should be no surprise to learn that this effort comes after the federal government threatened to cut off the city's community development funds if we did not proceed in this way. We are not unique: Chicago, Washington, and San Antonio, among others, have been put in the same position. Alameda, California, in fact, voted to reject the federal funds rather than undertake this effort. What may be surprising, however, is that the Black neighborhoods and elected officials have also strongly objected to this effort—not because they oppose increased housing opportunities for low-income and minority families and individuals, but rather because those same Section 8-assisted housing units are the prime resources for providing low-cost, quality rental housing in severely deteriorated minority communities. If these resources are not provided to those neighborhoods, deterioration will continue and eventually the families and individuals living there now will be forced to move because of present blight and hopelessness for the future.

Limited resources have drawn our objectives into conflict. The resolution of this conflict, as determined by the courts and administrative regulations, involves using 50 percent of the units for each objective. That is, a purely administrative uninformed compromise has been reached. Unfortunately, I have heard nothing at this conference that suggests that any other action might be considered more appropriate. In other cases, secondary effects of policies and programs raise relocation problems. Center-city urban renewal programs, intended at least in part to encourage middle-class mobility into inner cities, have resulted in involuntary moves of lower-income residents. Certainly, this is a problem that the research community has identified in the past. But, again, this problem is susceptible to more direct involvement, in that, after identifying a mobility-impaired group, a researcher might suggest resource commitments to ameliorate the problem. Other secondary effects have not been so clearly specified. The same highway improvements that facilitate continued movement to new suburbs by maintaining an abundance of accessible land also spawn suburban shopping centers. The success of these centers has contributed to the decline of inner-city commercial districts, thereby encouraging greater movement of residents to suburbs than would otherwise be the case. These resulting conflicts are beginning to be addressed in specific metropolitan areas, such as Dayton, and in national policy on federal assistance that supports shopping center development. However, the impetus for these efforts to modify legislation comes from local government self-interest, not from any true understanding of appropriate ways to control these patterns of relocation either in general or in specific cities.

Finally, the limits on our resources demand that we choose programs which achieve our distributional objectives efficiently. How much land at what level of accessibility is required to avoid overinflated prices? How high must interest rates become to substantially inhibit mobility? How efficient is affirmative action in housing programs in achieving housing opportunity compared with direct housing construction? In what circumstances? How extensively must homes be rehabilitated to stabilize neighborhoods? Is housing rehabilitation more efficient than new construction in stabilizing communities? The public sector has dozens of possible ways to achieve these objectives, but little understanding of how to choose among them—at least in terms of their varying consumption or distributional effects. Some research has been done to answer these questions—the housing allowance experiments and community development evaluation projects discussed at this conference are good examples—but not nearly enough is being undertaken. We need both general understandings of the efficiency of alternative programs as well as adaptions of those analyses to specific situations and localities. In the absence of such research, program choice resembles a trip to the department store—we buy what's in fashion. Currently fashionable items include housing rehabilitation, Section 8 housing certificates, and neighborhood commercial revitalization.

Second, the research community has the responsibility to consider the compatibility of public mechanics to the research findings that underlie recommended legislation. Most programs intended to achieve change through mobility are implemented by local governments, although they are often supported by national resources. National legislation is interpreted administratively for its "legislative intent," clarified through the courts, codified in administrative regulations, and operationalized in guidelines. These guidelines are further interpreted for a given locality by decentralized administrative units (area or regional offices) and eventually applied to segments of resources often allocated through administrative discretion. How often does the research community follow a program through these public mechanics to require that the results are consistent with the intellectual basis of the original legislative commitment? I suspect this seldom occurs—not because I have any direct evidence of such neglect, but rather because the programs I am asked to implement often seem to stray so far from the apparent original intents.

For instance, a few years ago, Philadelphia received a grant for improvements to a neighborhood commercial center, located within a ten-square-mile low-income district. These funds were provided under a federal program intended to provide public works employment in order to stabilize distressed communities which had suffered job losses. The administrative conditions accompanying the grant required that the workers reside on the few blocks immediately adjoining the center.

When it became clear that there were not enough prospective workers living on these few blocks, the city requested that the target employment area be expanded to include the entire low-income district. That suggestion was rejected, and the city was required to hire from the entire city area. There were two results: The intent of the program to stabilize the distressed community was totally lost, and local staff effort required to complete the project (which eventually took four years) was completely out of proportion to the benefits achieved.

Third, the research community has the responsibility, in recommending program reporting requirements, to determine that the data required are essential to monitor achievement of mobility objectives, not merely to establish compliance with guidelines. Collecting and reporting data on program performance is an enormously expensive activity. We spend hundreds of thousands of dollars annually in Philadelphia maintaining performance data on the Community Development Block Grant Program alone. When these data assist in local decision-making, their collection is usually worth the cost. This is particularly true when the data help with resource allocation. When the reporting grows out of the intergovernmental environment and the data collected merely establish compliance with guidelines that are not consistent with original policy objectives, their collection wastes staff time. Worse, such guidelines tend to change rapidly, resulting either in a change in reporting requirements or in the retention of requirements for which there is not even a procedural necessity. For three years, for instance, our office was required to collect data on the census tract of residence of employees of businesses we assisted. Collecting this data is quite difficult; but the program was intended to keep workers from having to move to other parts of the country to find jobs, not to keep them from moving to other census tracts.

If the intents of public policy relating to residential mobility are as durable as I have argued above, then it would seem that data required to determine the achievement of these objectives should also be consistent for long periods. The research community often is involved in suggesting program reporting requirements as part of federally supported program evaluation efforts. Researchers would do a significant service to local practitioners and to efficiency in program implementation by designing these reporting procedures to be consistent with the ultimate policy objectives.

It would be unfair to suggest that the research community has totally failed to address the questions necessary to fulfill these responsibilities. There are, obviously, instances where research efforts have helped with these issues. Furthermore, since many of these questions concern goal conflict, it might be reasonable to argue that these choices are public issues, not ones that are the responsibility of the research community to answer. However, my point here is somewhat different. Whether or not research findings exist to answer these questions, those findings have not been imbedded in the procedures of the public sector. The research community has been successful in achieving public policy response when the type of research necessary to effect that response involved the identification of problems and the suggestion of resource commitments necessary to attack those problems. In addition, the research community has been willing to and effective at expanding the set of public values concerning the role of mobility.

In short, the research community has been able to influence public policy. It has been willing to address questions of public values. The research community should now be willing to assume the additional responsibilities I have outlined here in order to assure that the resources committed to facilitate change through mobility are effectively used to achieve those objectives.

16

Data Resources for Monitoring Change

JANET W. BYLER and RANDOLF GSCHWIND

□ OVER THE LAST FEW YEARS there has been a notable change in the relation between federal programs and local implementation in the United States, changes which have a significant impact on the collection and use of data at the local level. The shift from a categorical to a block grant entitlement approach for allocating federal funds to housing and community development brought with it a shift in the locus of responsibility for program design and implementation from the federal to the local level of government. The regulations accompanying the Housing and Community Act of 1974, in particular, place primary responsibility on counties and municipalities for defining community needs and objectives and for formulating programs to meet them.

Implicit in the block grant approach of this legislation is the assumption that programs designed at the local level could be tailored to, and hence made more responsive to, changes in local conditions than the previous categorical programs. The validity of this assumption rests, in part, on the validity of two others: (1) that local policy makers have the capabilities and resources to formulate and evaluate alternative courses of action thoroughly and soundly, and (2) that a better knowledge of local problems and conditions exists at the local level. While recent work by Grigsby et al. (1977) has questioned the former assumption, this chapter seeks to address some of the issues raised by the latter–namely, the adequacy of existing data bases for planning at the local level. In particular, we are concerned with the ability to address policy issues which are sensitive to and impinge upon the characteristics of local residential mobility and neighborhood change. Our discussion will proceed in four parts.

In the first part, we provide a motivation for the development of accounting models of local change, which leads us to a discussion of collection and use of data about individual households and dwelling units in the next section. While most academic arguments for access to such micro-level data stress the importance of this sort of data in providing answers to certain classes of social questions (Gale, 1978; Juster, 1970; Orcutt, 1970), governmental use of this same data leans more toward operational and administrative uses. In our discussion, we argue that further payoffs to local government may lie in making available existing data in slightly enhanced forms to permit alternative kinds of analyses by their own planning staff as well as by social researchers. Then, in the third part, assuming that use of micro-level data files can help to improve our understanding of contemporary urban processes, we turn our attention to the problems surrounding the construction and manipulation of these files. In this part of our discussion we will be especially concerned with the abilities of local agencies to collect, organize, and disseminate micro-level data. Finally, we discuss the policy implications of micro-level file construction and access. To this end, we offer recommendations which could lead to more productive uses of this data without sacrificing concerns for privacy and confidentiality.

MOBILITY AND LOCAL ACCOUNTS

Given the decentralized nature of most data collection and statistical reporting in the United States, the assertion that knowledge of local problems and local conditions is better at the local than the federal level is in a certain sense tautologous. The real problem besetting those who attempt to formulate more workable programs for community development is not who knows more, but whether anyone knows enough and whether that knowledge can be put to use. The issue is not merely one of compiling bigger and more reliable local data bases, but of organizing such data to serve the dual purposes of informing both the decision-making function and our understanding of the interaction between more general processes and local context. However, the starting point for our argument lies not with the data but with existing theories of the operation of housing markets and the residential structures which result.

In theory, the link between the demand for a commodity and the available supply is provided by an economic market which functions primarily as an allocation mechanism. Unlike many other kinds of exchange processes, however, the market for housing is dominated by the use, reuse, and modification of secondhand capital assets and by the need to assess a wide variety of externalities (Grigsby, 1963: 21). Furthermore, each household is at once both a potential consumer of alternative units and a potential supplier of its own unit. Together with changes in the demographic, social, and economic conditions of households in the housing stock and in the physical environment of small areas, this duality of roles results in continuing short-term fluctuations in the composition of both the consuming population and the available dwelling units. In addition, these fluctuations are often distorted and even magnified by limitations in information on market conditions and by the effects of particular institutional constraints on transactions such as those arising from legal and administrative regulation.

Although the conventional microeconomic framework can be amended to cope with some of these properties of the exchange process for housing (see, for example, Weinberg et al., 1979), alternative conceptualizations based on accounting principles become increasingly attractive. Accounting models such as those suggested by White (1970, 1971), Hua (1972), Moore and Gale (1973), and Rees and Wilson (1975) treat the detailed sociospatial structure of the population more explicitly than conventional models, although they lack the latter's ability for data reduction. This orientation is particularly pertinent inasmuch as the demand for a wide variety of public services is a function not of sheer numbers but of the detailed composition of the population; decisions on the provision of such services require accurate estimates of the details of localized population shifts. Furthermore, owing to the potential speed of such processes as residential blight, effective intervention to halt or reverse this deterioration requires detailed, up-to-date information on a variety of socioeconomic and demographic characteristics of both households and dwelling units.

The logical structure of accounting models is developed in another paper (Byler and Gale, 1978). Here we are concerned with the ability of local communities to support such accounting-based models and their monitoring functions. Given that data on individuals, which form the foundation of these models, are extremely costly to gather on a citywide basis, it becomes essential that as much use as possible be made of existing data collected for a wide range of administrative as well as planning functions.

INFORMATION FOR LOCAL MANAGEMENT

Widespread recognition of the need to improve the information available for policy and decision-making led in the late sixties and early seventies to a number of proposals for statistical reform (Bauer, 1966; Sheldon and Moore, 1968; Land, 1975). The accompanying critiques of existing approaches to the generation and use of information focused on three issues: (1) the need to document more fully historical sequences of social change, (2) the importance of developing measures of social well-being to complement existing economic indicators, and (3) the potential enrichment of the data base which could result from the pooling of records from administrative as well as data collection agencies of various levels of government.

Much of this debate, though originally begun at the national level, quickly evolved into attempts to improve informational resources at the state, county, and municipal levels. The resulting efforts to develop urban indicators, to construct social accounts, and to create urban and regional data banks, though intellectually and conceptually distinct, shared a common weakness. In each case, little effort was made to reconcile procedures for identifying and collecting particular data items with consideration of their subsequent use in planning and policy contexts. As might be expected, the informational value of such data systems was generally low relative to their costs; as a consequence, most of these earlier large-scale efforts at statistical reform have been abandoned. Nevertheless, there remain problems of identifying the kinds of data actually needed and the procedures appropriate to their collection, storage, integration, and use to inform urban planning and policy-making.

Here we will address issues associated with the design of planning information systems for local governments. Information systems of the sort we will consider are an outgrowth of earlier efforts to create urban data banks and consist of a data base together with a set of ongoing methods and procedures for updating and using these data. Substantively, our interests lie in the design of planning information systems to monitor patterns of residential mobility and occupancy in urban areas and to support the analytic accounting functions outlined earlier.

Most existing local information systems have been designed primarily to use secondary data from federal, state, and private sources to characterize the attributes of the housing stock and population in urban areas. This reliance on external sources of data, however, imposes a number of restrictions on the kinds of data which are available to users at the local level. Typically, these limitations are both substantive (that is, in terms of the classes of phenomena which are reported) and structural (that is, with respect to the usual aggregate, cross-sectional nature of the data, the asynchronous periodicity of reporting for different sets of data, and so on [Holleb, 1969]). The impacts of these restrictions have been felt not only at the local level, but also within the larger policy community as the available data provide few reliable clues as to the short-run dynamics of housing markets, processes of residential mobility, or changing patterns of occupancy within urban areas. Micro-level data, data about the attributes of individual households, land parcels, dwelling structures, and housing units could provide the kind of information needed both to better characterize local conditions and to enhance our understanding of processes of urban change. Therefore, throughout our discussion we will focus on problems associated with the development of planning information systems designed primarily to generate, store, process, and retrieve micro-level data developed within local governments and which would be appropriate for monitoring residential mobility and occupancy at a high level of disaggregation.

In this section we will begin by describing what we mean by monitoring activities at the local level. Next, we will present an overview of the requirements for an information system to monitor residential mobility and occupancy using micro-level data. To this end, we will consider the availability, utility, and problems associated with the use of micro-level data at the local level. Then, in the following section, we will propose some possible solutions to these problems.

THE NATURE OF MONITORING

Monitoring, as we use the term, simply means keeping track of something. As we shall see, both the content and the structure of information systems reflect differences in the way we choose to define that "something" of interest. We can identify three separate, though interrelated, dimensions which shape that definition.

First, at the most basic level, we can differentiate between data bases and their associated information systems on substantive grounds in terms of the kinds of entities and classes of attributes or behavior being monitored. That is, some systems might be designed to monitor the tenure status or value of housing units, while others are used to monitor the income, size, or composition of households.

Differences in the entities and the attributes being recorded affect the scope of the kinds of questions the data base can address. However, their impact on the organizational structure of the required information system, especially if data collection for planning purposes is integrated with that of administrative agencies, is far from obvious. Insofar as organizational and institutional conditions vary from one jurisdiction to another, we will not attempt to provide a universal specification of the effects of monitoring different classes of attributes or different kinds of entities on the design of planning information systems.

Second, recognizing the hierarchies implicit in moving from individuals to households to communities or from housing units to structures to neighborhoods, we can differentiate between monitoring systems in terms of the level of disaggregation of the entities of interest. The chief distinction we will make on these grounds will be between micro-level data about the attributes and behavior of entities at the more disaggregate levels and macro-level data about the aggregate characteristics of groups of these entities.

Finally, we can differentiate between systems in terms of two alternative structural conceptions of monitoring. Under a stock conception, monitoring may be thought of as keeping track of the status of something either continuously or at particular points in time. Monitoring of this sort might, for example, be concerned with recording which households are occupying which dwelling units at particular points in time. Alternatively, under a flow conception, monitoring may be thought of as keeping track of patterns and interrelationships of changes in the status of things. Monitoring of this sort might record both the destination of outmigrants from an area and the origins of the households (if any) replacing them in a given set of dwelling units.

TYPES OF MONITORING SYSTEMS

Using these three dimensions we can identify a range of alternative kinds of activities and associated information systems for monitoring residential mobility and occupancy.

Macro-level systems. The least complex systems we can envision would simply record the status at a macro level of a single set of entities; new data would enter the system in an aggregate form, replacing existing information in the data base. Monitoring systems of this kind might, for example, keep track of current neighborhood or census-tract-level statistics for the housing stock in an urban area. The resultant data base would, of course, permit cross-sectional, but not historical, analyses of characteristics of the housing stock. A somewhat more complicated system would monitor changes in these neighborhood-level statistics over time; new data would still enter the system in an aggregate form, but existing data would be linked to this new information through a common set of macro-level identifiers (for example, census tract numbers). These linkages would permit the construction of historical sequences to support comparative static analyses of the characteristics of the housing stock. Still more complex systems might monitor changes in macro-level characteristics of two or more sets of entities. A system of this kind would require unique but consistent sets of identifiers for linking aggregate-level data across sets of entities (for example, associating socioeconomic and housing stock characteristics for a given census tract), as well as for linking data within a single set of entities over time. Nevertheless, the system would again be able to support only aggregate, cross-sectional, or comparative static analyses of patterns of occupancy and residential mobility.

Micro-level systems. The least complex system for monitoring at the micro level would simply record the current status of individual entities of interest; replacement of existing information with new data would permit, as in the macro-level case, only cross-sectional analyses. A system for this type of monitoring might keep track of the current attributes of individual dwelling units in an area. The chief advantage of this system over the corresponding macro-level systems lies in its ability to support the development of alternative sets of aggregations of the characteristics of the entities; that is, with appropriate sets of spatial descriptors, the system could provide not only census tract but also other planning district statistics, even in the case where these two sets of aggregations are not coincident but overlapping. More complex systems might monitor changes in the characteristics of micro-level entities; such longitudinal monitoring would require the use of unique but consistent identifiers for each individual from one observation to the next to link existing information with new data about specific individuals. As in the macro-level case, these linkages would permit the construction of historical sequences; this time, however, the patterns of association would permit processual as well as cross-sectional and comparative static analyses.

Finally, the most complex form of monitoring, keeping track of changes in the interrelationships among various sorts of micro-level entities, would require a system with complex sets of pointers or identifiers to link disaggregate characterizations across sets of entities as well as within single sets of entities. Such systems would record both historical and relational patterns among the sets of entities of interest, permitting the joint specification of existing patterns of attributes and behavior, as well as processual and historical analyses of sequences of micro-level changes and adjustments leading to these conditions.

Design issues. Given this range of alternative information systems, the problem of designing an appropriate system for monitoring at the local level is one of matching the informational needs of planners and policy makers with the available resources and capabilities. Planning activities, unlike most of the administrative and operational functions of local governments, for the most part are not directly concerned with using data at the micro level. Problems are defined and alternative courses of action developed and evaluated with little attention paid to their impact on specific households or dwelling units; nevertheless, micro-level data play an important role in meeting the informational needs of planners.

Most data, regardless of their destined use, are originally collected at the micro level; thus, although infrequently used at the individual level, micro-level data in some aggregate form are used at every step of the planning process to assess existing conditions, to develop praxeologic explanations, and to predict the consequences of alternative courses of action. Therefore, the relevant factor in developing new informational resources is not the level at which the data are used in planning, but whether existing macro-level data can provide the detailed understanding of local conditions needed to support planning and policy-making. Furthermore, insofar as planning attempts to understand and influence ongoing processes of urban change, it requires information not only about the current state and pattern of attributes and events, but also about the nature of these processes as well.

Most existing planning information systems were designed primarily to process, store, and use macro-level data to support cross-sectional or comparative static analysis. The notable exceptions to this, for the most part, use micro-level, address-based parcel files developed by a few planning departments. Typically, the on-line monitoring capabilities of these systems are limited to cross-sectional, not longitudinal, descriptions as new data are used to replace existing information in updating the data base (Hysom, 1973).

Basics of information system development. Despite the potential usefulness of micro-level, flow-oriented data systems to provide processual information about joint changes in the detailed distributions of the attributes and behavior of the population and housing stock of small areas, little effort has been devoted to creating systems of this sort. Advances in hardware and software design have greatly enhanced technical capabilities to generate, store, process, and retrieve micro-level data; there still exist, however, a number of barriers to their use. Perhaps the two most important problems are the lack of empirical justification for the use of micro-level modeling and the tentative nature of the results which are available. The crucial question to potential users of data of this sort is one of payoff for effort. At this time, the sophistication of modeling required for micro-level research often seems to outweigh its ability to provide answers to the questions facing local policy makers. The relative newness of micro-level theory construction and its rapid development also make for a lack of tried and true methodologies for local researchers to utilize when they may already be faced with skepticism toward "esoteric" quantitative methods.

In addition, misgivings over the potential utility of alternative sources of information, whether based on micro-level data or not, may be well-founded on organizational and institutional grounds. Knowing more, especially about the outcomes and impacts of controversial projects and policies, can be risky, legally and politically, for local decision makers. This is particularly true if the outcomes differ strongly in the wrong direction from those which were intended or if patterns of economic or racial discrimination emerge from subsequent evaluations and analyses. Similarly, evidence which goes against widely held assumptions about the nature of local conditions is not likely to be believed and the source of evidence impugned. Furthermore, under federal regulations, applications for federal (CDBG) funding for housing and community development require the use of data which are commonly available in all jurisdictions, effectively vitiating attempts at the local level to use complex, custom-designed micro-level data resources to support urban planning and policy-making.

Finally, despite recent changes in information technology, attempts to merge existing micro-level files from administrative agencies and other sources to provide micro-level data for planning purposes face a number of technical, conceptual, and legal problems. Although they may be kept in machine-readable form, the structure and content of most local data files still pertain to the original use for which the data were collected. As little thought was given in creating these files to potential uses in other areas, correspondingly little effort was made to develop common sets of identifiers and definitions for the entities and attributes being recorded by various agencies. Thus, most files are structured only to be useful for certain specific applications and to be used by specific applications-oriented software. In addition, many files created in the early days of local computerization (the early- to mid-sixties) remain in a form amenable to hardware and software available at that time; inertia and the cost of conversion dictate that many of these potentially usable files will retain their existing, though inefficient, structures for some time to come. Although most data are collected at the micro level, access to data at this level is frequently restricted even within the operating agencies of local governments. Efforts to ensure privacy and confidentiality of sensitive information are the primary reasons for the imposition of such restrictions. Pragmatic considerations, such as lack of funds or personnel, may impose additional practical restrictions on the availability of data which could otherwise be considered public information.

DATA SYSTEMS

We now consider ways of overcoming some of the barriers to the use of micro-level, flow-oriented data to monitor changes in population and housing market conditions. Essentially, the suggested strategy is to integrate the production of data for planning and management with the collection and generation of data by other functions of government. Clearly, this strategy is not without its own problems. Although much of the current record-keeping for administrative and operational control involves the classes of entities and kinds of attributes and behavior of interest in monitoring residential mobility and occupancy, differences in the way the data are intended to be used suggest that the problem of designing a planning information system based on these data files ought not to be treated simply as one of adapting or adding on to existing systems. Most existing systems, for instance, have been designed to promote efficiency in individual record retrieval and to replace existing information with current data as they become available. These systems are not intended to facilitate the linkage of records either through time or across files to depict patterns of existing conditions and events or to provide historical sequences of changes in these patterns. Therefore, we will emphasize the need to adopt a set of procedures for data-base administration to ensure definitional and structural comparability between files which are intended to provide data for planning as well as administrative and operational purposes.

Local governments generate or procure an abundance of micro-level data about individuals, households, dwelling units, residential structures, and land parcels from a variety of sources. For various reasons, some of these data may be of questionable utility for monitoring patterns and processes of change in population and housing characteristics. Some of the data, for instance, might be unreliable at the disaggregate level but still useful in more aggregate forms to provide an indication of changing patterns of association among events. Similarly, the long periodicity or even sporadic nature of the observations may render some sources of micro-level data less useful than might be the case with more systematic collection. In addition, incomplete coverage, biased sampling, and assorted definitional incomparabilities might diminish the potential utility of various sources of micro-level data. In spite of all this, much useful data are, or could be made, available to facilitate monitoring at the local level.

In this section, we will summarize some of the technical and conceptual issues involved in the development of planning information systems based on micro-level data (see Byler and Gschwind, 1979). As a practical matter, such systems must rely on computerized processing of data to facilitate storage, manipulation, and retrieval. The relationship between the development of such systems and earlier efforts to develop computerized urban data banks and municipal information systems is both subtle and complex. The experiences gained in these earlier efforts not only shape the kinds of files, the hardware, the software, and the organizational aspects of existing local data-processing systems, but also influence the way in which new systems might be created at the local level. In light of previous critiques of these earlier efforts (for example, Dunn, 1974; Kraemer et al., 1974), it is important to see our discussion as an attempt, in part, to address some of the issues which were neglected in previous efforts to develop urban data banks and information systems.

Earlier, we suggested that the principle benefits of using micro-level, flow-oriented information systems lay in their ability to support a variety of alternative classes of aggregations and to portray patterns of associations among sets of entities. The first trait is, of course, inherent in the nature of micro-level data. The second trait, however, must be built into the data base through the specification of

(1) a set of protocols and procedures for assuring the privacy and confidentiality of the data;
(2) compatible and consistent definitions for sets of entities, attributes, and other kinds of descriptors;
(3) one or more sets of unique yet consistent sets of identifiers for the entities of interest; and
(4) a set of system-specific conventions for storing, organizing, and accessing records and data files to facilitate linkage.

The first three issues are dealt with in depth elsewhere (Byler and Gschwind, 1979). Here it is sufficient to note that there are significant trade-offs between preserving confidentiality and retaining an ability to develop longitudinal record linkages which are critical to the analyses of mobility (Moore, 1978; Moore and Clatworthy, 1978). Planners, as we noted above, rarely use micro-level data in their original form, but at some aggregate level. For simple cross-sectional analyses based on aggregations from a single file, nonidentifiability within the data base of individual entities is of limited consequence. However, identifiability really becomes an issue when records from one file must be linked to corresponding records in another file. Then, whether the intended analyses are cross-sectional or longitudinal is itself of little consequence, for direct linkage of records at the micro level requires the use of unique or semi-unique identifiers for each of the entities in question.

The relation between identifiers and record linkage also arises in evaluating alternative spatial descriptors or geocodes. Clearly, if linkages are to be made on the basis of such codes, they must be unique; block or tract identifiers are insufficient. Existing technology such as the ADMATCH or UNIMATCH systems in association with DIME files can handle the linkage problems provided the base data are error-free, but does nothing to alleviate the confidentiality issue. Different strategies such as those discussed in Byler and Gschwind (1979) can provide partial solutions.

DATA ORGANIZATION AND STORAGE

Before proceeding, two definitions are in order. A "record" consists of a set of data about one or more characteristics of an entity; the values of attributes such as the size, structure type, and location of a specific dwelling unit at a given point in time, for example, may form a single record. A "file" consists of one or more records for any number of entities; continuing our example, in order to develop a complex file recording the occupancy structure of an area, records for dwelling units might be matched with one or more records for the individuals occupying these units through an identifier common to both sets of records (Byler, 1978, forthcoming).

Once relationships between the files and the records within the files are clarified, a choice must be made between two alternative ways of maintaining within the information system itself the patterns of association among related sets of records. The simpler of the two options is the creation of physically linked data files; the more flexible is the creation of logically linked files.

In physical linking, the separate sets of records for each entity are actually combined or at least placed in contiguous storage. Records for several sets of data for a single set of housing units might, for example, be formed into a single file by sorting the records in the source files on a consistent identifier or key such as address and then merging the files in sequence on the key. Multiple files can, of course, be linked pairwise by matching each item either on the same or on a different set of identifiers to further increase the complexity and, hence, the representational ability of the data base. Given appropriate sets of identifiers for the records, data for housing units might, for example, be merged with other data on the attributes of households occupying these units to provide information on the occupancy pattern of an area. Subsequent linkage on the household identifiers of a similar file for a later time period would begin to generate micro-level residential histories; linkage on the dwelling unit identifiers would produce corresponding occupancy histories of the units themselves.

Logical linkage of files through the use of associator tables or indexes to point to related records on physically disparate files provides an alternative to actual physical modes of linkage. Basically, this mode of organization works like the index of a book: References to certain items can be looked up in the index and then accessed via a map pointing to the appropriate storage locations. As with physical linkage, logical linkage can support associations among multiple sets of records; instead of actually merging the files, however, this mode of data storage and organization simply constructs more complex kinds of indexes for relating individual sets of records within the files. In effect, logical linkage of multiple files must, in other than a pairwise sense, be based on a scheme of master indexes for cross-referencing indexes of individual sets of files. Moreover, insofar as interest focuses on files exhibiting asymmetrical (that is, non-one-to-one) relationships among the contents of files, two separate tables of association must be developed, one relating records in the second to those in the first, and vice versa.

Most previous attempts to create planning information systems have relied on physical linkage of files for data organization and storage. The main advantages of this approach lie in the portability and transferability of the relational data generated in merging the files, the relatively low cost and requirements for machine storage to maintain the linkages, and the conceptual simplicity and ease of working with the physically merged data. Changes in available information technology, particularly the development of commercially supported data base management systems and the vastly reduced costs of machine storage, favor the creation of new systems based more on logical than physical modes of data organization. The main advantage of using a logical mode in the planning context lies in the fact that the physical file structures of the constituents of the data base often need not be appreciably altered. Not only can individual files retain their original format and organization, but, given appropriate communications capabilities, administrative or operational files which could be usefully related in the planning context can also be kept in physically remote locations. Moreover, because the links themselves may be separate from the data, this ability to use in-place files reduces the vulnerability of the information about relationships among records. In particular, because the linkage of files only takes place in the index, a significant potential threat to privacy can be avoided by restricting access to the index. Finally, logical modes of organization can support more direct and efficient forms of storage than is possible with many physical modes; therefore, although file indexes are expensive to build and maintain, some of the development costs can be recovered if frequent access to specific records is necessary.

The problem of choosing between alternative modes of data organization and storage depends, in part, on the intended use of the data base; on the technical capabilities of the available hardware, software, and data processing personnel; and on institutional arrangements governing access to the various data files of interest to planners. Regardless of the actual choice of a mode of organization, the structural schemes underlying the patterns of linkages, whether physical or logical, remain the same. However, we would like to stress that the choice between using one or the other of these schemes for organizing the internal structure of the data should not be made solely on technical grounds to facilitate operating efficiency of the system; instead, given the limited capacity of existing software to process hierarchical files of this sort, we feel that consideration of the intended use of the files should be of equal importance.

The final issue having to do with data storage and organization involves the design of files and procedures for updating the information in the data base. In a flow-oriented or longitudinally based system, updating files does not change the linkage structure, provided the keys remain constant. It will, however, change the way the linked entity looks—more information now exists on the entity, and some provision for handling the additional data must be made. Several possibilities include (1) creating new records as new observations are entered, indicating new status, but also retaining the old status record so comparisons can be made; (2) adding change information into fixed record fields originally created as filler space to accommodate new information; or (3) modifying existing records to reflect the new information by adding trailer segments or additional fields to existing records. The optimal method will depend on many factors, including available hardware and software, the overall data structure, how the data are to be analyzed, and how data on changes or events are collected and processed.

Some files—particularly accounting files—are updated continuously and in "real-time"; that is, at or near the time of the change. Generally, for research on residential mobility it is reasonable to examine change over some set time span (for example, a year) so that transactions can be accumulated and processed more cheaply at specified intervals. We should note, however, that recent changes in procedures of data capture, particularly those utilizing direct entry to produce continuous files, and the development of direct entry hard and software may increase the rate of updating and hence research aimed at examining change in much shorter time frames than a year.

LONGITUDINALLY LINKED FILES

Bringing all the concepts together leads us to a general file structure for residential mobility research using individual-level data. By linking different sets of micro-data to provide a richer data base, we can look at changes over a relatively short time period to go beyond cross-sectional analysis to analysis of processes of urban change (see Byler, 1978; Byler and Gale, 1978; Moore, 1978). Linking files over time provides perhaps the most formidable and time-consuming barrier to overcome; for not only do the characteristics of entities change, but another order of magnitude of links must be reconciled. In longitudinal monitoring, structures and characteristics may change, thereby greatly increasing problems of matching. Thus, it is all the more important that the capabilities for creating linked longitudinal micro-data files be designed into the system from the beginning.

Longitudinal files require that we incorporate several different levels of association in their design. One level involves linking of micro-data in different files pertaining to households or housing units separately. These links may change over time, as may the files which are being linked. At another level, the sets of files relating to housing units and households are linked together to provide more comprehensive data relating these entities. Finally, at a third level these entities are linked over time to show short-term changes in their attributes and in their relationship. Within any area, the pool of housing and the pool of households are constantly flowing, turning over and intermingling, with roughly 25 to 35 percent of all urban households changing housing units each year. Clearly, the ability to match dwelling units or households over time depends critically on decisions made in regard to the nature of identifiers. Usually this is not a problem for dwellings, but for households which move the lack of unique or consistent identifiers makes matching excessively expensive.

Much of the success in developing longitudinal files depends on informed data base administration. Unfortunately, many current efforts to develop information systems at the local level have ignored the more subtle problems of data base administration in favor of addressing those amenable to technological solutions. That is, instead of trying to ensure the use of common definitions and identifiers among the files used by various agencies, efforts have focused instead simply on getting disparate data sets out of hard copy files and into machine-readable form. The hope frequently evidenced by these efforts—that the use of sophisticated data base management systems will eventually overcome most of the problems associated with using files constructed for a diversity of other purposes—seems ill-founded in light of our previous discussion. What is needed, instead, is the recognition at the local level of the benefits to be realized by integrating the production of data for planning and policy-making with its collection and generation for purposes of administrative record-keeping and operational control. We would stress, moreover, that by integration we do not mean some post hoc use of files in place in other agencies, but a program of system design based on detailed assessments of the classes of data of potential use across various agencies.

POLICY IMPLICATIONS

Drawing back from the technical details, we conclude our discussion with an overview and brief analysis of the policy implications of the construction and use of information systems based on micro-level data. Throughout our discussion, we have explored from a technical perspective some of the issues confronting efforts to monitor change mobility and occupancy at the local level. In the previous section, for example, we examined some of the technical and conceptual prerequisites to file-matching. Much of that discussion involved the importance of the kinds of conventions and procedures used to capture, store, organize, and process the data of interest.

When approached from this perspective, the problem of information system design can be removed from a narrow focus on technical issues to a broader consideration of the organizational and institutional issues involved. Adoption of a program of data-base administration, for example, should be seen not simply as a move to promote technical compatibility and efficiency, but as an organizational necessity before attempts at building complex sets of interrelated files from different sources can succeed. Unfortunately, the long-standing tradition in the United States of decentralized data collection even at the local level does not bode well for the adoption of either a vigorous or far-reaching program of standardization. For data base administration, with its responsibility for building consistent and compatible files, implies a centralization of jurisdiction and control over the kinds of data which are available for administration, operational control, planning, policy-making, and, hence, a change in organizational authority.

On the other hand, in light of the increased availability at the local level of both hardware with ever-greater capacities and software packages with sophisticated data-handling capabilities, data base thinking is coming about naturally, but perhaps from the wrong direction. Local data base thinking tends to view the computers and the data base management software as the means to the integrated utilization of various sets of data. We might say that the technological and application systems usually dominate the form of the data. In true data base thinking, the opposite should occur (Martin, 1977); the data dominate the system. "Logical data base design" is an approach to the organization of data independent of the systems to be run and is aimed at providing information through maximum exploitation of logical relationships in the data. In most present local situations, the physical data base is usually dominant over the logical form.

Since information is the key ingredient to both local use of data for pragmatic decision-making and the more theoretical uses of data for enhancing understanding of urban processes, we can hope to see an increased acceptance of the use of logical data base design. This provides the type of micro-level, longitudinally linked files we are concerned with here as a natural by-product.

Information systems are no longer independent entities, as was often the case with file systems. They become highly interdependent through the relationships developed within and between data sets; thus, physical and logical redundancies of data are also eliminated. The type of data structure described leads much more readily to the use of what is known as "logical systems design" or "structural systems analysis." We suggest here that the natural progression of local data systems toward logical designs will at least provide the potential for better micro-level information sets, although access may still be limited; furthermore, the biggest problems in developing data base systems are not technical, but organizational and administrative.

Similarly, given the decentralized nature of data collection, the problem of securing access to potentially sensitive micro-level files raises a number of organizational, social, and political issues. Frequently, at the local level, the head of an agency responsible for collecting or generating a particular set of data is also responsible. for determining the conditions governing its release. Thus, such an organizational structure can work to restrict access even within the local governments themselves. Add to this the legitimate social and political concerns over the increased potential for invasions of privacy, and the outlook for support of systems based on micro-level data is not good.

In fact, eliciting the organizational and institutional support needed to construct micro-level-based systems will probably require pragmatic demonstrations of actual benefits which could accrue only with their use. Clearly, only the use of micro-level files would permit the detailed characterization of the housing and population of small areas or the identification and characterization of spatially dispersed target populations. Similarly, only the construction and use of longitudinal files would permit analyses of micro-level processes of change. With such data, then, analysts could not only provide more detailed and accurate descriptions of existing conditions, but also could begin to construct praxeologic explanations of the possible consequences of pursuing particular courses of action.

As the nature of overall data structures changes with the ability to create and process more elaborate data relationships, local use of the data also tends to change. Rising expectations for micro-level information on the part of local administrators and policy makers may speed up the inevitable migration to micro-level data base systems. Presently, more and more administrative use of detailed micro-level data is being made in a number of areas, particularly the following:

- community development planning,
- community and human service planning,
- housing assistance, planning
- housing inspection and rehabilitation,
- neighborhood planning,
- refuse collection and other utilities,

- school planning,
- transportation planning, and
- urban design.

Micro-data are most commonly used in these activities for work program planning, management, and evaluation. They permit decisions to be made in the light of continual change in those local communities. For example, one local government utilized micro-level land use, ownership, and tenure data for a specific residential target area to direct different mailings to resident owners, absentee owners, and tenants in single-family and duplex buildings, explaining a low-interest local loan program sponsored by the city. The program was attempting to stimulate reinvestment in the area and increase the level of owner occupancy. At the other end of the policy spectrum, attention is being directed to the ability of individual-level records over time to provide much more focused evaluations of the specific effects of government programs on certain classes of housing and households.

Probably the most important consequence which will stem from the local availability of accurate and timely micro-data is the new-found ability of government to perform strategic planning and tactical intervention in a much more immediate time frame and at a much smaller scale; namely, for the individual housing unit or household. Through closer government monitoring of changes at the individual level, urban programs and processes can be lifted out of constrictive and sometimes discriminating geopolitical boundaries and into, it is hoped, a more objective and effective mode of operation.

It is at this point that the importance of the interrelationships among the questions which arise in planning and policy-making, the ways we structure potential answers (theories) to those questions, and the kinds of data which are available become apparent. Existing theories and, to a large extent, the questions themselves have been shaped by existing aggregate, cross-sectional data resources and reinforced by efforts to collect more of the same sorts of data to test the hypotheses generated by these theories. And if processual or even disaggregate theory construction has lagged behind cross-sectional analyses, two unrelated changes in the planning environment might yet provide the impetus for micro-level file construction. First, experience with electronic processing of administrative and operational files is increasing both the potential data base and the level of expertise available at the local level. Second, when responsibility for program development and evaluation resided in federal authorities, planners and policy makers were, by and large, consumers of secondary data. With the shift in the locus of responsibility to the local level, local planners are in the position of producing as well as consuming the data needed for planning research and evaluation.

Given the structure of the regulations and reporting requirements for federally funded programs such as CDBG, whether planners will use micro-level data for monitoring residential mobility and occupancy is, in part, a matter for the federal government to decide. If the federal government continues, for example, to disallow use of locally developed data resources for supporting applications for funding, little incentive will be felt at the local level to continue the development of micro-level files. Similarly, if the reporting requirements for grantee performance under the CDBG program are lessened to avoid overburdening local governments, no incentive at all to develop reporting systems based on micro-level assessments of outcomes and program impacts will be felt at the local level. Thus, federal policy will itself continue to play an important role in shaping the informational resources available for planning at the local level.

REFERENCES

BAUER, R. A. [ed.] (1966) Social Indicators. Cambridge, MA: MIT Press.

BIDERMAN, A. D. (1966a) "Anticipatory studies and standby research capabilities," pp. 272-301 in R. A. Bauer (ed.) Social Indicators. Cambridge, MA: MIT Press.

--- (1966b) "Social indicators and goals," pp. 68-153 in R. A. Bauer (ed.) Social Indicators. Cambridge, MA: MIT Press.

BYLER, J. W. (1980) "Information for urban decisionmakers: some conceptual issues in the design of urban information systems." Doctoral dissertation, Graduate Group in Peace Science, University of Pennsylvania. (unpublished)

--- (1978) "A dual flow conception of occupancy pattern change." Presented at the 74th Annual Meetings of the Association of American Geographers, New Orleans.

——— (forthcoming) Information for Urban Policy Makers: Some Conceptual Issues in the Design of Urban Information Systems. Doctoral dissertation, University of Pennsylvania.

——— and S. GALE (1978) "Social accounts and planning for changes in urban housing markets." Environment and Planning A 10: 247-266.

BYLER, J. W. and R. A. GSCHWIND (1979) "Local monitoring of residential mobility and occupancy." CSDE draft paper. Philadelphia: School of Public and Urban Policy, University of Pennsylvania.

City of Milwaukee (1977) P.D.I.S. Section, Central Electronic Data Services. Master Property User's Guide.

DIAL, O. E. (1968) Urban Information Systems: A Bibliographic Essay. Cambridge, MA: Urban Systems Laboratory, Massachusetts Institute of Technology.

DATTEN, A. R. (1977) "The city of Toronto central property register." Papers from the 15th Annual Meeting of the Urban and Regional Information Systems Association 3: 123ff.

DOWNS, A. (1976) Urban Problems and Prospects. Chicago: Rand McNally.

DUNN, E. S., Jr. (1974) Social Information Processing and Statistical Systems: Change and Reform. New York: John Wiley.

——— (1972) "The national data bank movement in the United States." Proceedings of the American Statistical Association, Business and Economic Statistics Section.

——— (1971) "The national economic accounts: a case study of the evolution toward integrated statistical information systems." Survey of Current Business 51: 45-64.

——— (1967) "The idea of a national data center and the issue of personal privacy." American Statistician 21: 21-27.

GALE, S. (1978) "Remarks on information needs for the study of geographic mobility," pp. 13-48 in W.A.V. Clark and E. G. Moore (eds.) Population Mobility and Geographic Change. Studies in Geography 25. Evanston, IL: Northwestern University.

GARN, H. A. and M. J. FLAX (1971) "Urban institute indicator program." Working paper 12061. Washington, DC: The Urban Institute.

GLASER, E., E. D. ROSENBLATT, and M. K. WOOD (1967) "The design of a federal statistical data center." American Statistician 21: 12-20.

GRIGSBY, W. G. (1963) Housing Markets and Public Policy. Philadelphia: University of Pennsylvania Press.

——— et al. (1977) Rethinking Housing and Community Development Policy. Philadelphia: Department of City and Regional Planning, University of Pennsylvania.

GROSS, B. M. (1966) "The state of the nation: social systems accounting," pp. 154-271 in R. A. Bauer (ed.) Social Indicators. Cambridge, MA: MIT Press.

HOLLEB, D. B. (1969) Social and Economic Information for Urban Planning. Chicago: Center for Urban Studies, University of Chicago.

HUA, C. I. (1972) Modeling Housing Vacancy Transfer in the Study of Housing Sector Interaction. Ph.D. dissertation, Harvard University. (unpublished)

HYSOM, J. L. (1973) "The urban development information system–a land use decisionmaking tool in Fairfax County, VA." Papers from the 11th Annual Meeting of the Urban and Regional Information Systems Association: 43-57.

——— W. B. RUCKER, N. AHUJA, and R. P. AHNER (1974) A Handbook Creating an Urban Development Information System. Fairfax, VA: County of Fairfax.

JUSTER, F. T. (1970) "Microdata, economic research, and the production of economic knowledge." Papers and Proceedings, American Economic Association, 60: 138-148.

KRAEMER, K. L. et al. (1974) Integrated Municipal Information Systems: The Use of the Computer in Local Government. New York: Praeger.

LAND, K. C. (1975) "Social indicator models: an overview," pp. 5-36 in K. C. Land and S. Spilerman (eds.) Social Indicator Models. New York: Russell Sage.

--- and S. SPILERMAN [eds.] (1975) Social Indicator Models. New York: Russell Sage.

LEVY, N. P. (1978) "An information system for city and neighborhood planning." Papers from the 16th Annual Meeting of the Urban and Regional Information System Association 1: 238-247.

MARTIN, J. (1977) Computer Data-Base Organization. Englewood Cliffs, NJ: Prentice-Hall.

MOORE, E. G. (1978) "The impact of residential mobility on population characteristics at the neighborhood level," pp. 151-181 in W.A.V. Clark and E. G. Moore (eds.) Population Mobility and Residential Change. Studies in Geography 25. Evanston, IL: Northwestern University.

--- and S. J. CLATWORTHY (1978) "The role of urban data systems in the analysis of housing issues," pp. 228-258 in L. S. Bourne and J. R. Hitchcock (eds.) Urban Housing Markets: Recent Directions in Research and Policy. Toronto: University of Toronto Press.

MOORE, E. G. and S. GALE (1973) "Comments on models of occupancy patterns and neighborhood change," pp. 135-173 in E. G. Moore (ed.) Models of Residential Location and Relocation in the City. Evanston, IL: Northwestern University Press.

ORCUTT, G. H. (1970) "Basic data for policy and public decisions: technical aspects." Papers and Proceedings, American Economic Association 60: 2.

REES, P. H. and A. G. WILSON (1975) "A comparison of available models of population change." Regional Studies 9: 39-61.

RUGGLES, N. and R. RUGGLES (1970) The Design of Economic Accounts. New York: National Bureau of Economic Research General Series, Columbia University Press.

STRUYK, R. J. (1979) "The need for local flexibility in U.S. housing policy." Policy Analysis 3: 471-483.

SHELDON, E. B. and W. E. MOORE (1968) Indicators of Social Change. New York: Russell Sage.

SYMONS, D. C. (1974) "An urban information system." Proceedings of the Fourth European Symposium on Urban Data Management, Madrid.

WEINBERG, D. H., J. FRIEDMAN, and S. K. MAYO (1979) "A disequilibrium model of housing search and residential mobility." Presented at a conference on Housing Choices of Low Income Families, Washington, D.C.

WHITE, H. C. (1971) "Multipliers, vacancy chains, and filtering in housing." Journal of the American Institute of Planners 37: 88-94.

--- (1970) Chains of Opportunity. Cambridge, MA: Harvard University Press.

17

Continuing the Debate

W.A.V. CLARK and ERIC G. MOORE

☐ IN THIS FINAL ESSAY we provide an overview of the main issues arising out of the discussions at the conference and indicate those areas into which future policy-oriented research might be primarily directed. Throughout, it should be remembered that we are concerned with the links between mobility research and public policy. Certainly, we recognize that research need not be policy-relevant; however, there are many social scientists who contribute to public policy as consultants, expert witnesses, or public critics. It behooves them to make sure not only that their research addresses problems of public concern but that it is structured in such a way as to be sensitive to the nature of public decision-making. To be effective requires both a focus on implementation, given the available instruments of public policy, and on communication between academic concerns and public policy interests.

Most conference participants agreed that mobility in and of itself is of little interest to public policy. Studies describing individual decision processes or aggregate patterns of movement do little more than provide background information. It is the specific outcomes of mobility which must be linked to issues of public concern. These outcomes can be defined at a number of different levels. Interregional distribution of population is a basic component of changes in the social fabric of the nation. The movement to the suburbs and increasing concentration of minorities in central cities continues to transform the character of our urban areas. Processes of neighborhood change, residential segregation, and integration alter both the residential environment of individual families and the distribution of demand for a wide range of public services. Individual household relocations are a basic mechanism for adjusting housing consumption, and any programs designed to affect the latter must consider the role played by mobility. Finally, the outcomes at all of these levels create a more general and pervasive dynamic, which results in continual variation in the local contexts in which policies are designed, implemented, and evaluated.

The chapters in this volume provide specific research contributions focusing on each of the issues described above except for interregional population redistribution. However, discussion tended to focus on four more general issues: the distinctions between academic research and policy concerns, the role played by models, the nature of data needs for policy-related mobility research, and the social context of mobility. It is those four issues which are used to organize the following comments.

LINKS BETWEEN RESEARCH AND POLICY

There are at least three different constituencies relevant to our discussion: academic social scientists, individuals concerned with the design and implementation of policies at the federal level, and those involved with these tasks at the local level. The prime reason for drawing the distinction between federal and local concerns is that the former necessarily demands a much stronger focus on generalization across cities and hence lies close to the traditional academic concern for identifying and understanding consistent relationships in society, while the latter is more pragmatically concerned with problems of allocation within specific communities. Since the financial support of research is almost entirely confined to the federal (and state) levels, the result is a considerable alienation of local governments whose interests are often ill-served by the research process.

In contrasting the behavior of these constituencies, an important distinction is made between general research on social processes favored by universities and targeted research required by government agencies. It is clear that although both political and budgetary constraints demand a focus on specific problems of public interest, too narrow a focus will often lead to system-wide effects of public actions being ignored. The frequent result of a narrow focus is that unintended outcomes are generated. Kevin McCarthy of the Rand Corporation emphasized this point in terms of the Department of Housing and Urban Development's current mission-oriented research. He sees two types of programs being developed. The first focuses on low-income households and the need to provide housing of a reasonable standard at a reasonable cost. In such programs, the primary focus is on changing household consumption. The second type of program focuses on improving the conditions in specific neighborhoods or communities through investment in rehabilitation of the housing stock, in facilities, infrastructure, and public works. Such programs are place-oriented. This double focus can produce tensions. On the one hand, an open-housing policy is intended to promote increased access and mobility and to stimulate minority access to white neighborhoods; on the other hand, neighborhood improvement is designed to discourage abandonment and flight, thus decreasing mobility. In addition, the frequent result of households competing for more desirable locations is that investment in poorer neighborhoods leads to lower-income households being forced out as local conditions improve. Such a redistributive outcome runs counter to those policy goals of improving dwellings and neighborhoods for those with limited incomes.

There are additional problems which are concerned more specifically with policy mechanics. The policy implications of research are often presented as general statements without any attempt to interpret them within the framework of local social and economic constraints. For example, it is one thing to say that maintenance of the racial composition of integrated housing requires replacement of white households during normal processes of occupancy turnover, but quite another to specify what types of management practices will encourage that replacement. We cannot dismiss the latter as a purely administrative problem. Since there is considerable variation in the size, location, and composition of integrated housing projects, we would expect management practices to interact with local circumstances. Unravelling these interactions requires considerable analytic skills–skills which reside primarily in the academic community.

In focusing on policy mechanics it is important to remember that instruments of policy usually control supply rather than demand. Limitations on new construction, redevelopment and rehabilitation, investment in infrastructure, and transportation improvements all provide examples of supply-side influences. In contrast, most models of mobility and neighborhood change are driven by household demand (see, for example, the review by Quigley and Weinberg, 1977). If models of mobility are to be made more effective, they must include not only instrumental variables but also a linkage of supply and demand-side variables–possibly, as Mark Menchik suggests, through the development of simultaneous equation models.

THE ROLE OF MODELS

While there is no question that a wide variety of models ranging from straightforward accounting models to complex simulations play a critical role in academic research, their function in the policy arena cannot be so clearly defined. From the federal perspective, some degree of modeling seems essential in order to make sense out of the sizable differences in mobility, housing adjustment, and neighborhood change among metropolitan areas when these metropolitan areas exhibit considerable variation in local economies, political organization, demographic structure, and housing stock (Struyk, 1979). However, at the local level the demands on models are different. Not only are issues of generalizability less critical, but problems tend to be resolved in a more immediate public forum in which simple powers of persuasion dominate. To cope with this environment, models must be simple, robust, and capable of being understood by the various participants in the decision-making process.

Ginsberg made an eloquent argument for focusing much greater attention on the ability of various actors to organize and utilize different types of information. This comment applies to both modeling efforts and data base development. Each mode of organization, whether it be a private business in a market context, a bureaucratic agency, or a neighborhood group, has its own information requirements which are suited to the ways in which decisions are made in that organization. Highly centralized organizations have a need to synthesize information in order to make high-level strategic decisions. At this level, the use of elaborate models is encouraged. For organizational decisions which are more decentralized, the specifics of time and place dominate and the need for analytic models as opposed to up-to-date description diminishes.

From this perspective, academic researchers also represent an organizational entity. We might argue that sophisticated models are an integral part of their pursuit of knowledge. One of the main thrusts of Ginsberg's argument, however, is that analysis of public policy is not the same as the design and implementation of public policy. The tools useful to one are not necessarily useful to the other. Unfortunately, we know little about the effectiveness of different types of models in influencing decisions in the policy domain, although recent debate suggests quite strongly that expectations–particularly about the role of large, complex models–have been overly optimistic (Batty, 1979; Pack, 1978; Sayer, 1979).

The critical point is that models which are used to gain insight into social processes cannot simply be translated into policy by demonstrating a general relationship between these social processes and problems of public concern. We must address not only the question of policy mechanics but also the way in which the outcomes (of the modeling effort) can be communicated to those responsible for making decisions.

INFORMATION NEEDS

Until quite recently, studies of mobility had been dependent on two sources of data: the decennial census and cross-sectional surveys designed to answer specific research questions. As a consequence, we learned a great deal about the variance in the propensity to move for households with different compositions, but almost nothing about the nature of adjustment in housing consumption over time, nor about the rate at which such adjustments produced shifts in the character of urban neighborhoods. From a policy perspective, these latter gaps are critical. The Experimental Housing Assistance Program (EHAP) funded by the Department of Housing and Urban Development, as well as the Seattle and Denver Income Maintenance Experiments, has made a considerable contribution in terms of producing longitudinal data appropriate to the analysis of housing adjustments in different program contexts. However, these programs have produced strong indications that specific responses are also a function of local market conditions (Hanushek and Quigley, 1979). This finding, coupled with the basic need to document changes in local conditions, places a strong pressure on local communities to develop their own monitoring capabilities. If the cost of monitoring is to be kept within reasonable limits, it is imperative that as much use as possible be made of existing administrative files, a task which depends on a number of quite technical issues.

The development of data bases for both urban management and policy analysis also raises questions similar to those in the previous section. In the same way that different organizational entities have varying responses to models, so will they have different data needs. Again, Ginsberg developed the argument that sophisticated data base management systems are most appropriate for highly structured tasks and have not proven useful when unstructured, exploratory problem-solving situations are the norm. He felt that the latter situation was the most common in local planning and policy-making. This is debatable, for one might argue that there are a large number of decision-support situations, particularly those relating to detailed changes in the characteristics of local populations, which require consistent and replicable procedures for collecting, organizing, and analyzing data. Even if the latter position can be maintained, the general point that different organizations have different data needs is crucial to future system development.

THE SOCIETAL CONTEXT

Both in the individual papers and in the discussions there have been strong suggestions that the decision to move must be regarded as but one of a number of related decisions necessary to achieve household goals. An emphasis on independent decisions would also lead to a demand for closer links between research on housing and research on metropolitan labor markets. Mobility is played out in the contexts of where people live, the dwellings they occupy, where they work, and the facilities they use. However, it is only in the outcomes of mobility that moving becomes of major significance for policy. As Lowry noted in the first section of this book, the changes in urban form and its specific representation in terms of residential integration or lack of integration exemplifies the translation of movement to a basic societal problem. To analyze the problem of the lack of residential integration without considering mobility is as short-sighted as analyzing mobility without considering the social outcomes. These mixed-mode analyses are still lacking in social science approaches, and it is only the development of broader societal perspectives as distinct from disciplinary perspectives which will yield more general and policy-relevant results.

More specifically, we know that the general set of social relationships and constraints are the background against which we analyze any problem of concern to social scientists. For example, the ways in which wealth is distributed and redistributed, the limits on public spending and allocation provide a powerful set of constraints on the actions of individuals and groups. The challenge for this macro-level analysis lies in setting the concepts of individual choice with a broad theory of society (Duncan, 1976; Sayer, 1979). Only to the extent that we can see individual actions and outcomes in a societal context will we be able to understand the complex effects of public policy actions.

A FINAL COMMENT

The natural instability of public policy goals arising from the vagaries of the political process complicates any attempt to identify and stimulate policy-relevant analysis. In this sense, the research on residential mobility is no different from that in the other social sciences. Policy analysts in the federal and local bureaucracies often have backgrounds in the social sciences, but they have to deal with problems which do not fit readily within specific disciplines. As we have seen several times in this volume, the fragmentation of research on mobility across several disciplines is perceived as a serious barrier to the integration of research and policy.

The broader issue which arises out of this specific situation and from the concerns with the role of models and of data relates to the way in which individuals who enter the public policy field are trained. In most universities policy analysis is still dominated by disciplinary interests and the traditional academic concerns. If professional training is to be provided in the policy field, then to some extent that training must be separated from existing disciplinary structures. In this way the training may be an effective bridge between the motivations, goals, and instruments of both research and policy. Questions of substance, such as those relating to mobility and neighborhood change, are then more likely to be responsive to the constraints and realities of the public policy arena. The nature of the training and the continuing link between social scientists and the practitioners of public policy may in the end be the most critical connection between substantive research and policy decisions.

REFERENCES

BATTY, M. (1979) "Progress, success and failure in urban modelling." Environment and Planning, A 11: 863-878.

DUNCAN, S. S. (1976) "Research directions in social geography: housing opportunities and constraints." Transactions of the Institute of British Geographers, New Series 1: 10-19.

HANUSHEK, E. A. and J. M. QUIGLEY (1979) Complex Public Subsidies and Complex Household Behavior: Consumption Aspects of Housing Allowances. Working Paper 825. New Haven, CT: Institution for Social and Policy Studies, Yale University.

PACK, J. R. (1978) Urban Models: Diffusion and Policy Application. Monograph Series No. 7. Philadelphia: Regional Science Research Institute.

QUIGLEY, J. M. and D. H. WEINBERG (1977) "Intra-urban residential mobility: a review and synthesis." International Regional Science Review 3: 41-66.

SAYER, R. A. (1979) "Understanding urban models versus understanding cities." Environment and Planning, A 11: 853-862.

STRUYK, R. J. (1979) "The need for local flexibility in U.S. housing Policy." Policy Analysis 3: 471-483.

The Contributors

JANET W. BYLER is Research Associate in the Community Development Strategies Evaluation in the School of Public and Urban Policy at the University of Pennsylvania. Her current research focuses on the design of urban information systems for the development and evaluation of public policy both at the local and the federal level.

W.A.V. CLARK is Professor of Geography and Associate Director of the Institute for Social Science Research at the University of California, Los Angeles. His research interests are focused on intraurban migration and neighborhood change. He is co-author of *Los Angeles: The Metropolitan Experience* (1976) and co-editor of *Population Mobility and Residential Change* (1978).

MICHAEL DEAR is Associate Professor of Geography at McMaster University in Hamilton, Ontario. He is editor (with A. J. Scott) of *Urbanization and Urban Planning in Capitalist Society*.

NORMAN I. FAINSTEIN is Associate Professor of Urban Affairs and Policy Analysis at the New School for Social Research. He is a senior researcher in the Community Development Strategies Evaluation project at the School of Public and Urban Policy at the University of Pennsylvania. With Susan Fainstein he has published in the general area of urban political economy.

SUSAN S. FAINSTEIN is Professor of Urban Planning and Policy Development at Livingston College, Rutgers University.

She is a senior researcher in the Community Development Strategies Evaluation project at the School of Public and Urban Policy and the University of Pennsylvania. With Norman Fainstein, she has published in the general area of urban political economy.

STEPHEN GALE is currently Chairman of the Regional Science Department at the University of Pennsylvania and co-Principal Investigator of the HUD-sponsored evaluation of the Community Development Block Grant Program. He is co-editor of *The Manipulated City* (1975).

MARTIN E. GOLDSMITH is presently Director of Community Development Planning for the City of Chicago.

RANDOLF GSCHWIND is Senior Analyst in the Department of City Development, City of Milwaukee, Wisconsin, focusing on the development of information systems for local planning and policy.

JAMES E. HARTLING is Deputy Director, Office of Housing and Community Development, City of Philadelphia. He previously taught community and regional planning at the University of Texas, Austin. His professional interests are in community development and urban economic development.

JAMES W. HUGHES is Professor of Urban and Regional Planning, Department of Urban Planning and Policy Development, Rutgers University. He has written extensively on urban planning and metropolitan change, his most recent books being *Post-Industrial America: Metropolitan Decline and Inter-Regional Job Shifts* (1978) and *Housing: Problems and Prospects* (1980).

JAMES S. LEMONIDES is with the City of Chicago's Department of Housing, Division of Research and Program Development.

JAMES T. LITTLE is Associate Professor of Economics at Washington University in St. Louis. His research interests

include economic theory and the nature of housing markets. He is co-author of *Neighborhood Change* (1975).

IRA S. LOWRY is at the Rand Corporation in Santa Monica, California, where he currently directs research on the HUD-sponsored Housing Assistance Supply Experiment. His publications include *A Model of Metropolis, Migration and Metropolitan Growth,* and "The Dismal Future of Central Cities."

MARK DAVID MENCHIK conducts public policy research at the Rand Corporation. In addition to residential mobility, he is currently investigating the fiscal limitation movement and the effects of cities' economic decline on their labor forces.

WILLIAM MICHELSON is currently teaching in the Program in Social Ecology, University of California, Irvine. Recent books include *Environmental Choice, Human Behavior, and Residential Satisfaction* (Oxford University Press, 1977) *Public Policy in Temporal Perspective* (Mouton).

ERIC G. MOORE is Professor of Geography at Queen's University, Kingston, Ontario, and a senior researcher focusing on relocation impacts in the Community Development Strategies Evaluation project at the University of Pennsylvania. He is co-editor of *The Manipulated City* (1975) and *Population Mobility and Residential Change* (1978).

JOHN M. QUIGLEY is Professor of Public Policy at the University of California, Berkeley. His research interests include urban housing and labor markets and local public finance. He is co-author of *Housing Markets and Racial Discrimination: A Micro-Economic Analysis* (1975).

TERENCE R. SMITH is Associate Professor in the Department of Geography, University of California, Santa Barbara. His research interests center on individual decision-making and information-processing.

DANIEL H. WEINBERG is a senior economist at Abt Associates, Inc. His research interests include housing economics, local public finance, and welfare programs. He is co-author of *The Housing Choices of Low Income Families.*

ALAN G. WILSON, formerly Director of the Centre for Environmental Studies, is now Professor of Geography at the University of Leeds. Among his many books on urban and regional modeling are *Models of Cities and Regions* (1977) and *Spatial Population Analysis* (1977).

DATE DUE

GAYLORD PRINTED IN U.S.A.